Local Mediterranean Food Plants and Nutraceuticals

Forum of Nutrition

Vol. 59

Series Editor

Ibrahim Elmadfa *Vienna*

KARGER

Local Mediterranean Food Plants and Nutraceuticals

Volume Editors

Michael Heinrich *London*
Walter E. Müller *Frankfurt*
Claudio Galli *Milan*

24 figures and 23 tables, 2006

KARGER

Basel · Freiburg · Paris · London · New York ·
Bangalore · Bangkok · Singapore · Tokyo · Sydney

Michael Heinrich
Centre for Pharmacognosy and Phytotherapy
The School of Pharmacy
University of London
London, UK WC1N 1AX

Library of Congress Cataloging-in-Publication Data

Local Mediterranean food plants and nutraceuticals / volume editors, Michael Heinrich, Walter E. Müller, Claudio Galli.
p. ; cm. – (Forum of nutrition, ISSN 1660-0347 ; v. 59)
Includes bibliographical references and index.
ISBN-13: 978–3–8055–8124–0 (hardcover : alk. paper)
ISBN-10: 3–8055–8124–6 (hardcover : alk. paper)
1. Wild plants, Edible–Mediterranean Region. 2. Vegetables in human nutrition–Mediterranean Region. 3. Diet therapy–Mediterranean Region. 4. Ethnobotany–Mediterranean Region. I. Heinrich, Michael, 1957 July 4- II. Müller, Walter E. (Walter Erhard), 1947- III. Galli, Claudio. IV. Series.
[DNLM: 1. Nutritive Value–Mediterranean Region. 2. Plants, Edible–Mediterranean Region. 3. Diet Therapy–Mediterranean Region. 4. Diet, Mediterranean–Mediterranean Region. 5. Ethnobotany–methods–Mediterranean Region. W1 BI422 v.59 2006 / WB 430 L811 2006]
QP144.V44L63 2006
581.6′32–dc22

2006017134

Bibliographic Indices. This publication is listed in bibliographic services, including Current Contents® and Index Medicus.

Disclaimer. The statements, options and data contained in this publication are solely those of the individual authors and contributors and not of the publisher and the editor(s). The appearance of advertisements in the book is not a warranty, endorsement, or approval of the products or services advertised or of their effectiveness, quality or safety. The publisher and the editor(s) disclaim responsibility for any injury to persons or property resulting from any ideas, methods, instructions or products referred to in the content or advertisements.

Drug Dosage. The authors and the publisher have exerted every effort to ensure that drug selection and dosage set forth in this text are in accord with current recommendations and practice at the time of publication. However, in view of ongoing research, changes in government regulations, and the constant flow of information relating to drug therapy and drug reactions, the reader is urged to check the package insert for each drug for any change in indications and dosage and for added warnings and precautions. This is particularly important when the recommended agent is a new and/or infrequently employed drug.

www.karger.com
Printed in Switzerland on acid-free paper by Reinhardt Druck, Basel
ISSN 1660–0347
ISBN-10: 3–8055–8124–6
ISBN-13: 978–3–8055–8124–0

Contents

General Introduction

Over the millennia, human nutrition has developed from a hunter-gatherer one to one based on agriculture. The former obviously dictated the formation of small migrating communities, whose diet consisted predominantly of products from wild animals and wild plants, while the development of agriculture made larger communities possible, and allowed a diet based mainly on products from plants cultivated on a relatively large scale, and on the limited consumption of meat from flocks and herds.

Today, we depend on a tiny number of domesticated species and this is due to plant and animal domestication which is regarded to be the most important cultural development in the past 13,000 years of human history [1]. In many rural regions, especially of Southern and Central/Eastern Europe, non-cultivated food plants are still gathered or grown on a small scale and consumed as healthy 'snacks', salads, vegetables or are produced on a small scale giving rise to local varieties or cultivars. Such food has been called 'local food' [2].

In general terms, wild varieties tend to be richer in micronutrients and bioactive secondary metabolites than the corresponding cultivated ones. Such secondary plant metabolites are produced in adaptation to local environmental conditions, which in the Mediterranean area are often pro-oxidative. This would require adaptive responses by producing 'protective', bioactive compounds. It is of interest, in this respect, that polyphenolic molecules are produced in response to stress, and these compounds have been shown to activate a group of components, the sirtuin family of deacetylases, that had arisen in primordial eukaryotes, possibly to help them to cope with adverse conditions [3]. Sirtuins are found in plants, yeast and animals where they appear to be involved in slowing down the aging process, implying an underlying conserved mechanism. The transfer of compounds produced in response to stress signaling molecules from one species to another in their environment is the basis of the 'xenohormesis hypothesis' [3]. In this way, organisms can prepare in advance for a deteriorating environment.

Whatever the conceptual basis of the above hypothesis, it is very relevant for human health that the intake of minor compounds produced by plants in response to stress *results in the transfer of the protective effects to our organism*. Therefore, research in this area provides clues for improving our understanding of the mechanisms that induce the production of bioactive molecules in plants and in setting up strategies for the exploitation of their potential applications to optimized human nutrition.

Consequently, 'local foods' represent a type of mutual interactions between the availability of locally growing, edible plants, on one side, and the nutritional requirements and needs of populations living in those areas. The persistence of few of these 'spots' up to modern times provides the opportunity for investigating, with updated experimental approaches, the features of these plants, the nature of their nutrients and their potential role in health promotion.

Ethnobotany

Core to the project, which is summarized in this volume, is the ethnobotanical study of local food plants in selected regions of the Mediterranean. Ethnobotany investigates the relationship between humans and plants in all its complexity, and is generally based on a detailed observation and study of the use a society makes of plants, including all the beliefs and cultural practices associated with this use. Ethnobotanists live with the members of a community, share their everyday life and, of course, respect the cultures which host them. Ethnobotanists have a responsibility both to the scientific community as well as to the indigenous cultures. Ethnobotanists use a complex set of methods derived from the social and cultural sciences including the taking of detailed field notes and collect carefully documented plant samples (voucher specimens) that allow for precise determination of the plant species. Ethnobotanical studies have a multitude of theoretical and applied interests and in fact only very few are in any way directly linked with projects in the area of discovering novel food or pharmaceutical products. While many researchers in the natural sciences and the general public often see such ethnobotanical information as a source of inspiration for 'us', the continued knowledge about and use of these resources in the regions of origin will require not only their recognition as local knowledge or traditional ecological knowledge, but also their study and development from a multidisciplinary perspective. Clearly, this is only possible if the 'traditional keepers' of this knowledge have a say in its future use and benefit from such research and development.

Specifically, in this project the focus is on local Mediterranean food plants, a topic which despite the diversity of traditional food knowledge around the Mediterranean has until recently received relatively little attention in the scientific literature: wild gathered fruits and vegetables [cf. 1, 4–6].

Core to our approach is an emphasis on two closely linked but conceptually and methodologically distinct goals: On the one hand the ethnobotanical studies for the basis of developing leads for new nutraceuticals by characterizing plant extracts derived from plant species with potential health beneficial effects traditionally used in rural communities of Southern Italy, Greece (Crete) and Southern Spain (see especially chapters 4–7). As importantly, the project gives new value to local food products which have been used for many generations and which now are on the brink of becoming forgotten and seeks methods to assure the preservation of this knowledge for future generations by highlighting the local relevance of these resources (see especially chapter 3), by disseminating the information on a local level in the national languages [e.g. 4, 5]. Next, biological-pharmacological effects of selected plants were studied (see below).

Local Plant Foods and the Cardiovascular System

The first biological system that is encountered by food components in entering our body, after leaving the gastrointestinal tract, is the cardiovascular (CV) system. Throughout the whole life period, macro- and micronutrients and a variety of other compounds, most of them still largely undefined, are ingested, and maximal concentrations are reached in the post-prandial phase, a time period that is progressively attracting greater attention in research on the impact of nutrition on the CV system [7] and health in general. After digestion, absorption, possible metabolism and transport, nutrients also interact via the CV system with peripheral tissues. Of special relevance, with reference to the direct effects on CV health, are those exerted by compounds present in local food plants, and have been shown to act as modulators of vascular functions. Among them, flavonoids, comprising mainly quercetin and kaempferol, and a variety of phenolic compounds are able to exert risk-lowering effects on mortality from coronary heart disease [8].

A long-standing presence in the Mediterranean dietary habits is also covered by special foods or components of the diet such as olives and olive oil, and wine. Olive oil has been used throughout the millennia by numerous civilizations and presumably before the start of agriculture. To the peoples of the Mediterranean this oil has been more than a mere food: it has medicinal and magical uses and is an endless source of fascination and wonder. The phenolic components of olive oil have been shown to exert a variety of effects on modulators of the vascular system, both in vitro as well as in animal models and humans [9, 10]. Wine has also been a constant component of the diet in countries in the Mediterranean basin, as the millennial history of its use in ancient civilizations such as those in Greece, Israel and Rome shows. Wine, especially red wine, has been shown to contain microcomponents of phenolic nature that are potent antioxidants and exert protective actions on vessel walls, so that moderate consumption of this beverage appears to be a healthy habit.

Cellular constituents of vessel walls, from the endothelium to cells of the muscular layer and those that penetrate into the wall from blood, modulate vascular functions through the production of a variety of 'functional' modulators. These include on one side promoters of vasodilation (e.g. nitric oxide and prostacyclin), as well as those that are involved in 'inflammatory' responses, e.g. the cytokines and various types of proinflammatory molecules. In addition, the formation of oxidative derivatives of circulating macromolecules, such as oxidized lipoproteins, oxidative alterations of cellular DNA, contribute to the onset and progression of chronic pathologies in the CV system, such as the atherosclerotic disease. Such processes were shown in the studies described in this volume, devoted to the effects on the CV system, to be favorably affected by the phenolic components extracted from the plants under investigation.

Mediterranean Diet(s): Food for Thought?

Neurodegenerative events, due to neuronal loss or malfunction, include both normal brain aging and severe maladies such as Alzheimer's disease. Cell death within the brain, especially of neurons, is particularly troublesome as their new formation only proceeds at marginal levels. Recently, research advances in the understanding of the effect of fruits and vegetables rich in secondary plant metabolites – which for a long time have been considered as antinutrients – on animal and human physiology (especially of the brain), is gaining momentum [11].

Among all cellular tissues, the brain most tightly controls the access of organic and inorganic molecules, including dietary constituents and their metabolites, in order to minimize the impact of potentially hazardous molecules. Polyphenolic molecules, which have been shown to enter the brain in small amounts, are of particular interest, as they boost cellular resistance to deleterious oxidative stress due to a pleiotropic interplay with direct and indirect antioxidant mechanisms. Whereas the latter effects become evident by changes in the amplitude of biomarkers, for example, lipid peroxidations, numerous animal studies mirror the beneficial impact of plant foods and their constituents on brain function by means of improved performance in behavioral tests. Last but not least, large long-term epidemiological studies indicate that people greatly benefit in terms of reduced risk for the onset of severe neurodegeneration, by consuming a Mediterranean-style diet rich in fruits and vegetables [12].

These data, together with those obtained in studies on other biological systems [13], provide the background for applications to the development of nutritional strategies for health promotion and are an essential element of the overall approach presented in this volume. While over the last decades some information on such major food components has been gathered, prior to this project practically nothing had been known about local food.

This volume is the outcome of a European Union funded project on local components of the various variants of Mediterranean diets and will hopefully stimulate further ethnobotanical, pharmacological and nutritional studies in this emerging field [13]. We gratefully acknowledge the support of all partners of the consortium 'Local Food-Nutraceuticals' and, of course, the funding provided by the European Union (KA1 'Food, Nutrition and Health', FP5, QLRT-2001-00173; 2002–2004).

Michael Heinrich, London
Walter E. Müller, Frankfurt
Claudio Galli, Milan
May 2006

References

1 Leonti M, Nebel S, Rivera D, Heinrich M: Wild gathered food in the European Mediterranean: a comparison. Econ Bot 2006;60(2):(in press).
2 Heinrich M, Leonti M, Nebel S, Peschel W: 'Local food – nutraceuticals': an example of a multidisciplinary research project on local knowledge. J Pharmacol Physiol 2005;56(suppl):5–22.
3 International Conference on the Healthy Effect of Virgin Olive Oil: Consensus Report. Eur J Clin Invest 2005;35:421–424.
4 Lamming DW, Wood JG, Sinclair DA: Small molecules that regulate lifespan: evidence for xenohormesis. Mol Microbiol 2004;53:1003–1009.
5 Lau FC, Shukitt-Hale B, Joseph JA: The beneficial effects of fruit polyphenols on brain aging. Neurobiol Aging 2005;26(suppl 1):128–132.
6 Nebel S: Ta chòrta: Piante commestibili tradizionali di Gallicianò. London, University of London, School of Pharmacy, 2005, pp 1–61.
7 Rivera D, Obon C, Inocencio C, Heinrich M, Verde A, et al: The ethnobotanical study of local Mediterranean food plants as medicinal resources in southern Spain. J Physiol Pharmacol 2005;56(suppl):97–114.
8 Scarmeas N, Stern Y, Tang MX, Mayeux R, Luchsinger JA: Mediterranean diet and risk for Alzheimer's disease. Ann Neurol 2006 (DOI 10.1002/ana.20854).

9 Sies H, Schewe T, Steffen Y, Heis C, Kelm M: Flavonoids: mechanisms and cardiovascular health. Free Radic Res 2005;39(suppl 1):S22–S22.
10 Spencer JD: Importance of diet in vascular prevention: vastly underestimated. Circulation 2002;106:133.
11 Tardio J, Pardo de Santayana M, Morales R: Ethnobotanical review of wild edible plants in Spain. Bot J Linn Soc 2006 (in press).
12 The Local Food-Nutraceutical Consortium: Local Mediterranean diets: a joint ethnobotanical and phyto-pharmacological approach. Pharmacol Res 2005;52:353–366.
13 Visioli F, Galli C: Biological properties of olive oil phytochemicals. Crit Rev Food Sci Nutr 2002;42:209–221.

Heinrich M, Müller WE, Galli C (eds): Local Mediterranean Food Plants and Nutraceuticals. Forum Nutr. Basel, Karger, 2006, vol 59, pp 1–17

'Local Food-Nutraceuticals': Bridging the Gap between Local Knowledge and Global Needs

Michael Heinrich[a], *Sabine Nebel*[a], *Marco Leonti*[a], *Diego Rivera*[b], *Concepción Obón*[c]

[a]Centre for Pharmacognosy and Phytotherapy, The School of Pharmacy, University of London, London, UK; [b]Departamento de Biología Vegetal, Facultad de Biología, Universidad de Murcia, Murcia, [c]Departamento de Biología Aplicada, EPSO, Universidad Miguel Hernández, Orihuela, Alicante, Spain

Abstract

Food use is changing very fast all over the world. This and other changes (e.g. reduced physical activity, increased longevity) result in novel health risks for the populations in European countries and beyond. Also, in recent decades the convenience food market has grown dramatically and offers novel opportunities for small and large industries alike. Simultaneously, there is a dramatic and irrevocable loss of the local knowledge which forms the basis of many cultural traditions (traditional food knowledge – TFK). The Mediterranean region is well known for a dietary tradition commonly called 'Mediterranean diet(s)', which is renowned for health benefits based among others on widely consumed foods and beverages. While the focus of research has mostly been on the more widely used elements of the Mediterranean diets (especially olive oil and red wine), in this review the focus is on 'local food'. These are ingredients, which are gathered, grown or produced locally and prepared into dishes, which often represent local specialities. Such food is derived from animals, fungi and plants, but in this paper the main subject is food of botanical origin. Particularly important among these local foods are vegetables and salads derived from wild greens (gathered food plants – GFPs) and local cultivars of fruit trees and shrubs. In this review we discuss the theoretical basis (including the concept of traditional knowledge systems) and general approach of an EU-funded multidisciplinary ethnobotanical-pharmacological project focusing on the use of such local resources in several regions of the Mediterranean including the ethnobotanical documentation of food products of selected communities in southern Italy, Spain, Greece (mostly Crete), the identification of extracts/pure compounds (leads for new health food supplements) with potent activity on a series of in vitro targets, especially ones relevant to assess for antioxidant activity, the more detailed in vivo study of some lead extracts and lastly the dissemination of such TFK in local/national languages.

Introduction: The Changing Food Pattern

Food use is changing very fast on all continents. In industrialized countries there is a dramatic rise in expenditure for convenience food (pre-prepared or ready-made dishes) and at the same time nutraceuticals, functional foods and other food products with some health claim are a rapidly growing segment of the market. The latter is specifically in response to risk factors that impact on the health of everyone. Specifically, a series of risks are highlighted in the numerous studies focusing on diet and health, including:

- Overconsumption of energy (calories).
- High intake of glucose.
- Low intake of fruits and vegetables.
- Low intake of fiber.
- High intake of red meat.
- An imbalance in the ratio of ω3:6 fatty acids.

Aside from dietary aspects, other problems, such as environmental factors and lack of activity, have been implicated. This increases the risk of numerous diseases including various forms of cancer [1], cardiovascular [e.g. 2] and skeletomuscular diseases, but also numerous metabolic syndromes like diabetes. Taking diabetes as an example, it is now apparent that not only people in many developed countries suffer from the various forms of diabetes and the illnesses associated with it, but in the last decades the disease has also become a serious health problem in indigenous and local communities all over the world, including Mexico, the USA, Canada, Australia, but also in many regions of Asia and Polynesia [3–8]. Clearly, there is a direct link between changes in dietary traditions and the overall consumption of the various classes of foods on the one hand and the increased prevalence of this syndrome on the other. According to the World Health Organization [9], diabetes affects more than 176 million people worldwide. The changes in diet are difficult to evaluate since in many instances the 'traditional' or local forms of eating have disappeared and normally no quantitative data on food consumption in traditional societies is available. For many policymakers it seems such knowledge only becomes a topic of research once it has disappeared and health and sociocultural problems develop.

This review offers a critical overview on one specific system of knowledge – the Mediterranean region with a diverse and relatively well-known dietary tradition commonly called 'Mediterranean diet' [10] and specifically on local traditions which show a great degree of diversity. The nutritionist Ancel Keys found that firemen in Naples and Madrid had a significantly lower blood cholesterol level than Americans, and that this correlated with a lower percentage of certain types of fat in their daily diets with a predominance of mono- and polyunsaturated fatty acids (e.g. olive oil). Keys and collaborators [11, 12] also showed that major cultural differences exist in the distribution of risk characteristics and risk behaviors, including diet, and in the geographic and time distribution of the major causes of death. Keys demonstrated that traits, which previously had been considered irrevocable and constitutional, such as body type, blood fat levels (cholesterol), blood pressure, heart rate, and responses to stress, were, in fact, largely modifiable by simple changes in the composition and quantity of diet and physical activity and that 'traditional Mediterranean diet' helps to improve health parameters. In the next chapter, the focus will be on a discussion of what 'local knowledge' and more specifically 'local food knowledge' is.

Local Food as an Element of Local Knowledge

Knowledge and bearers of knowledge stand at the center of knowledge systems, which can be defined as 'the way people understand the world, and interpret and apply meaning to their experiences' [13]. Such knowledge is built through the complex process of selecting, rejecting, creating, and transforming information, and is inextricably linked to the social, environmental, economical and institutional contexts in which it occurs [14]. Numerous terms have been used for it and in the following we focus on two which are particularly relevant in the context of this volume: local knowledge and traditional (ecological) knowledge (TK or TEK).

The example chosen is the knowledge about food and how this knowledge impacts on the utilization of botanical resources in some regions of the Mediterranean. Even though the different social science disciplines have all contributed to an understanding of the social aspects of food consumption, hardly any discipline matches anthropology in contributing to our understanding of food consumption, perhaps the most classical and cited example being the culinary triangle of Lévi-Strauss [15].

Local traditions rely on information being passed on from one generation to the next in one community or in a small region. Food may form a part of local knowledge if its use is confined to a certain area for which it is a characteristic cultural trait. Traditional food knowledge (TFK) as part of TK is strongly influenced and determined by socioeconomic and cultural determinants, religion and history. The concept of local knowledge has been essential in the field of anthropology. Originally developed in the context of cultural anthropology in the USA, it is now recognized as an important theoretical vision of how to interpret knowledge systems, which often have not been recognized formally. Crucial have been the contributions by Clifford Geertz [see 16]. His collection of essays 'Local Knowledge – Further Essays in Interpretative Anthropology' clearly is a milestone in the discussion about how people interact with each other and their environment. C. Geertz is one of the co-founders of symbolic or interpretative anthropology, which developed in the late 1960s. Culture is – according to Geertz – a complex system of symbols, which allows humans to give meaning to the experiences they face and to communicate these experiences among each other. This culture-specific system of symbols should be recorded as a 'thick description', a term he had introduced into the field in 1973 (borrowing it from the British philosopher G. Ryle). It is essential to understand these systems of symbols and to properly describe them in detail. Thus a 'thick description' is one based on the experience of a researcher who often comes from a foreign culture, which simultaneously shows meaning systems which have not been visible to the indigenous groups. The symbolic role of 'local knowledge' is essential to Geertz's thought and has had a strong impact on anthropological thought.

In first instance it is crucial to point out that 'local knowledge' is an all-encompassing concept. As Geertz [16, p 167] has put it focusing on legal questions: 'Like sailing, gardening, politics, and poetry, law and ethnography are crafts of place: they work by the light of local knowledge'. Thus all aspects of culture rely on a complex system of thought which in its totality forms local knowledge. Therefore, local food knowledge or any other aspect of local knowledge is just one part of this complex system, which has an important meaning to the members of the culture. The challenge is to see '. . . the broad principles in parochial facts. "Wisdom", as an African proverb has it, "comes out of an ant heap". . .' [16, p 167].

These concepts do provide a strong theoretical foundation for working on the complex meaning systems of indigenous and 'traditional' societies. However, *they clearly lack a systematic and detailed discussion of applied aspects of such knowledge systems.* There is no input from applied researchers or from anyone who is interested in preserving, disseminating or using this information. Therefore, it is essential to highlight another stream of thought, which does come specifically from such an angle – the concept of *Traditional Ecological Knowledge*, which is also well known under its acronym TEK. This has been defined as 'a body of knowledge built up by a group of people through generations of living in close contact with nature' [17]. Another definition describes TEK as a 'cumulative body of knowledge, practice and belief evolving by adaptive processes and handed down through generations by cultural transmission, about the relationship of living beings (including humans) with one another and with their environment' [18]. Both definitions recognize that TEK is an attribute of societies in general with historical continuity in resource use practice. Interest in TEK has been growing in recent years, partly due to the recognition that such knowledge can contribute to the conservation of biodiversity [19], rare species, and to sustainable use in general [20].

It includes a system of classification, a set of empirical observations about the local environment, a system of self-management that covers resource use.

The quantity and quality of TEK varies among community members, depending upon gender, age, social status, intellectual capability and profession (hunter, spiritual leader, healer, etc.). With its roots firmly in the past, TEK is both cumulative and dynamic, building upon the experience of earlier generations and adapting to the new technological and socioeconomic changes of the present [17]. Compared to Geertz's emphasis, TEK highlights the applied role of such knowledge system and is little interested in symbolic or other theoretical aspects.

The concept is very popular with activists and applied researchers in all fields, which involve the local and sustainable development of resources, the critique of outside involvement in such developments and most notably the implications of the Convention on Biological Diversity (CBD or Convention of Rio de Janeiro, 1992) [21]. Despite its at first glance apparent limitations to ecological knowledge it normally encompasses botanical, zoological, geological, pharmaceutical and many other aspects of local knowledge. Many people now recognize the enormous contribution that TEK systems have made and can make in the future both to the conservation and the sustainable use of biological diversity. Most indigenous and local communities live in regions where the vast majority of the world's plant genetic resources are found. Simultaneously these activists and scholars argue for urgent action to safeguard such TEK and for a broad view of how humans and the environment interact. Importantly, this discussion proceeded with a focus on non-European indigenous groups while far less attention has been paid to the situation within Europe.

The origin of TEK are based on the knowledge that societies existed under conditions of constant pressure on the resources upon which they depended, and that a means had to be found to persuade communities and families to economize with regard to their use of natural resources [22].

TEK is a constantly evolving way of thinking about the world. Although views covered by TEK are described as 'traditional', this does not mean that they do not change or that it would be desirable that they do not adapt to new needs. Each generation makes their own observations, and compares their experiences with what they have been taught [23]. The practice of TEK differs from that of scientific ecological knowledge in that it is largely dependent on local social mechanisms [18]. Social mechanisms often play a role in the integration of ecological knowledge of different kinds. Mechanisms for the intergenerational transmission of knowledge are embedded in social systems. Successful transmission of agricultural skills and knowledge, for example, depends on the amount of time families spend on the land because of apprenticeship-based knowledge transmission, and the amount of time required for hands-on learning.

One of the important outcomes of the discussions on TEK systems has been the broad recognition, *that a new relationship between researchers and 'keepers of traditional knowledge' is needed.* Since it was agreed upon in 1992, the CBD has been at the center of the discussions about equitable benefit sharing between the 'South' and the 'North'. Some activists group argue for a complete ban on bio-prospecting or other research activities which have the potential of resulting in new economically exploitable products, while others argue for a novel relationship between keepers of TEK and (mostly 'Western') researchers (see later section 'Benefit Sharing').

In this review we do not want to summarize this complex and fast-moving discussion. However, the answers will depend on the region or country of origin of such potential new products and especially with respect to Europe a renewed interest in some aspects of TEK has resulted in local and regional economic opportunities for (mostly rural) populations.

These are examples from a diversity of human activities as they relate to human's use of the environment. In the next section, we highlight this link specifically as it relates to the use of local food in the Mediterranean basin.

Local Food – A Historical and Anthropological Perspective

Food is an example which exemplifies local knowledge or TEK. Plants, fungi and animals that yield food are one of the most essential resources for all humans and of course an everyday need. At the same time, in many regions food is still produced or gathered in the immediate surroundings of the house or the community. Food is a very basic need, but – unless people suffer from starvation – also a pleasant experience. Food and dishes do always reflect a 'vision of the world' and consequently peoples, ethnic groups and communities are proud of their special dishes and the plants or breeds of animal they produce and use. Such knowledge gives them a regional identity and of course is also the source of ethnic stereotypes. For example, the regular consumption of acorns as a source of proteins and carbohydrates has given rise to a pejorative term used to refer to the locals from the province of Badajoz in Spain ('Belloteros' = acorn eaters). However, it is well beyond the scope of this short overview to discuss the concept of 'local food' on a worldwide scale.

The Mediterranean TFK has sometimes been described as a diet mainly composed of pasta (in southern Italy), olive oil, vegetables (leafy vegetables, legumes and unprocessed cereals), fruits, red wine, seafoods and only few red meats and was termed the 'Mediterranean diet'. More precisely, Mediterranean TFK should be referred to as 'Mediterranean *diets*' since the many different cultures, religious beliefs, ecologic backgrounds and historic developments around the Mediterranean basin resulted in many diets, which share a multitude of elements, but also revolve around distinct local or regional traditions.

From an anthropological point of view, food culture can be defined as a culinary order whose traits are prevalent among a certain group of people [24]. The most basic distinction in the area of local nutrition is the categorization from a social point of view of *what is edible and what is not*. However, other distinctions such as criteria of (pleasant or unpleasant) taste, the relationship between certain food items and certain sociocultural contexts (ways of consumption, timing of meals and eating situations), and the values attached to food also contribute to the diversity of a food culture. The more specific term *cuisine* is often used to denote special typical ingredients, combinations of ingredients and preparation methods belonging to a certain ethnic group, region or country [25].

Clearly, local food plants have been an element of Mediterranean dietary traditions from the beginning of human's occupation of the region. This is beautifully expressed and contextualized in Cervantes' Don Quixote, written at the beginning of the 17th century AD [26]. The meager diet of the knight – wild dry fruits and meadow greens (thus *gathered food plants* – GFPs) – is presented as a decadent prototype of the medieval chivalry and seen by his squire as not substantial.

Chapter X: 'Pardon me, your worship,' said Sancho, 'for, as I cannot read or write, as I said just now, I neither know nor comprehend the rules of the profession of chivalry: henceforward I will stock the alforjas with every kind of dry fruit for your worship, as you are a knight; and for myself, as I am not one, I will furnish them with poultry and other things more substantial.'

'I do not say, Sancho,' replied Don Quixote, 'that it is imperative on knights-errant not to eat anything else but the fruits thou speakest of; only that their more usual diet must be those, and certain herbs they found in the fields which they knew and I know too.'

Clearly, they were often seen as a diet, which offers few or no attractions. Don Quixote himself found this diet so unpleasant that instead he preferred a piece of bread.

Chapter XVIII: 'In that case we have nothing to eat today,' replied Don Quixote. 'It would be so,' answered Sancho, 'if there were none of the herbs your worship says you know in these meadows, those with which knights-errant as unlucky as your worship are wont to supply such-like shortcomings.' 'For all that,' answered Don Quixote, 'I would rather have just now a quarter of bread, or a loaf and a couple of pilchards' heads, than all the herbs described by Dioscorides, even with Doctor Laguna's notes.'

Even more so, very often they were and are seen as the diet of the poor, the unrefined or ignorant. Don Quixote considered the regular consumption of GFPs to be a sort of penitence or mortification.

Chapter XXV: 'Let not that anxiety trouble thee,' replied Don Quixote, 'for even if I had it I should not eat anything but the herbs and the fruits which this meadow and these trees may yield me; the beauty of this business of mine lies in not eating, and in performing other mortifications.'

In a way this is a classical example of contrasting concepts of the upper, allegedly educated class (Don Quixote who behaves like a city-dweller or citoyen) and the lower, illiterate class (Sancho, who is a peasant). This conflict between the class that derives products from the fields and the one that gets it through trade is found in historical and modern examples. Moreover, this reflects the false perception a city-dweller (Cervantes himself) had of the rural life in a period of cosmopolitism and globalization when the Habsburg dynasty ruled most of Europe. The above talks were supposed to be kept between two 'manchegos' (inhabitants of the province Castilla-La Mancha). Both are native to a country where today over 60% of rural population knows and regularly consumes GFPs or 'meadow herbs' like 'collejas' [*Silene vulgaris* (Moench) Garcke] as substantial part of their diet [27, 28].

Of course and as exemplified in the above example, history shapes food cultures [29]. Another core element is the availability: traditional cuisines are obviously very dependent on the natural resources available in the area (see next part 'Gathered Food Plants and the Local Mediterranean Food'). Due to the importance of local geographical conditions for the availability of food items, regional patterns have traditionally been particularly relevant for food products. Such regional patterns persist today even if modern distribution systems have liberated local eating patterns from the constraints of climate. This is also exemplified in the slow changes of food habits in immigrant societies [29, cf. 30].

As a consequence one has to ask how does one define the concept of '*local* food'? In a geographical approximation, this clearly relates to ingredients, which are gathered, grown or produced locally and prepared into dishes, which often represent a local speciality. Such food is derived from animals, fungi and plants, but in the context of this discussion the focus is on food of botanical origin. Some of these local traditions have become worldwide phenomena (in a way a 'global food') but this binary opposition does not really help in explaining the differences between what is only known locally with more widespread types of food. In its core all these local foods have their origin in small-scale activities in a politically, culturally or socially well-defined region. If one were to create a binary opposition, the alternative to local food would be *food, which is commercialized well beyond the region where it is produced and which is derived from commercial large-scale production.*

Globalization has resulted in an increasing availability of similar foods in different cultures. Local food knowledge from Naples, Italy, was transformed into a worldwide food, now often considered to be a typical American dish. The pizza was brought over to the USA and on its onward journey back to many European countries it became an element of American cultures (even though without the discovery of the Americas the Aztecan 'tomatl' would not have found its way onto the Napolitanean pizza) and is now a prime example of a convenience food. This incorporation of new food species is, of course, not a new phenomenon, Turkish corn (Fuchs' Türkisch Korn) is included in many of the early European herbals like Fuchs, Brunswick and others. In fact, it refers to *Zea mays*, which, of course, is of Mesoamerican origin. Similarly, sweet and hot peppers (*Capsicum annuum* and *C. frutescens*) are now considered typical elements of the Hungarian and Balkan (as well as the Asian-Indian) cuisine, but are again of American origin. A telling example comes from the Mediterranean. In recent years, rocket salad (*Eruca sativa*, Brassicaceae) has become a new, popular food in certain sectors of the European Community. Some health benefits have been investigated scientifically [31]. The aerial parts were originally well known only in the Central provinces of Italy (especially Umbria) and it was an important food of the Ancient Romans, that later fell into disuse in most regions. Remnants of this widespread reputation in ancient times are the persisting local uses as flavoring agent in traditional dishes such as *gazpacho* of North Murcia and East Albacete provinces in southern Spain. Its leaves have a prominent, spicy taste and in recent years have become common ingredients of many salads in North and Central European countries. Thus, rocket salad is a prime example of the transformation of a local food into a commercial product.

Another example was highlighted in our research in southern Italy. Horseradish (*Armoracia rusticana*) is commonly consumed as a condiment, most notably during the carnival period in spring. Ground horseradish is used to aromatize a sauce used with homemade noodles, or omelettes consumed during Carnival. It is a fascinating semicultivated species in the community of Castelmezzano [32]. Semicultivated or naturalized horseradish is very abundant in the northern and Alpine areas of Italy but extremely rare in southern Italy. In Castelmezzano, this local food is commonly 'tolerated' in home gardens and local people insist that it has always been part of the local cuisine. It is likely that horseradish came to Castelmezzano (and to a few other villages of inland Lucania) via migrants from Swabia in the 13th century.

Thus as a theoretical summary, local food may well have its origin outside of the regions of use and the crucial aspect of its definition is the local production, gathering or harvesting and consumption.

Gathered Food Plants and the Local Mediterranean Food

So far the focus was on the local production, but ecological aspects form an important basis for understanding it. Local food is based on plants grown locally by individual peasants or gathered[1], for example, by shepherds and certain segments of the population or on local varieties of animals. Such food is

[1]From a nutritional perspective, edible fungi, while taxonomically different, are an essential element of these gathered foods.

often only available seasonally and is consumed either fresh (e.g. spring salads and vegetables, fruits in autumn) or in a conserved form (dried, fermented, pickled). Local may be defined from a sociocultural perspective [33] or within an ecological framework, discussing food production from the perspective of a basic ecological unit defined by its climate, soil, watershed, species and local agro-ecosystems. Two types are recognized on ecological and agronomic grounds: GFPs (weedy, synanthropic and wild species) and local cultivars. GFPs grow in primary or secondary habitats. Some are strictly endemics, others are widespread species [34]. Sociocultural factors determine the roles of GFPs and their importance is largely based on taste, palatability, toxicity, cultural taboos and difficulties in production or collection.

Some less appreciated local food plants are consumed as seasonal hunger food and as famine food. These have been used particularly in times of hunger or during seasonal chronic shortages. They are not consumed regularly due to factors such as accessibility, palatability or toxicity. Often these foods are culturally associated with marginalized classes or peoples from culturally and economically deprived areas. The 'forgotten crops' or 'secondary food crops' which were replaced with introduced species or cultivars considered to be more interesting generally also fall into this low-profile category.

Some high-profile Mediterranean GFPs are highly sought, and have reached a notable level of recognition in fine cuisine at local level. The reasons are not always concurrent with biological characteristics. In general they are palatable, low in toxicity, and accessibility is not a limiting factor in the appreciation of these foods, at most it influences their market value. The genus *Eruca*, now sold on a global level in large supermarkets (see above), and truffles, *Tuber nigrum*, highly appreciated in the world of top cuisine, are examples of such high-profile Mediterranean GFPs but very few others have become popular on a global scale.

The 'Local Food-Nutraceuticals' Project and Its Goals

Such local traditions were the focus of the project 'Local Food-Nutraceuticals' (LFN), which is summarized in the following paragraphs and which forms the basis for most of the chapters in this volume of *Forum of Nutrition*. The LFN project includes specialists in many fields like ethnobotany, pharmacognosy, pharmacology and nutritional studies, which come from the UK, Germany, Poland, Spain, Italy, Switzerland and Greece. This project is unique in the framework of the EU's programmes in being the first with a very *strong focus on ethnobotanical questions and in being truly multidisciplinary bridging the gap between sociocultural and natural sciences*. As outlined above, the interest in local foods stems only in part from its potential as a lead source for novel nutraceuticals. Such food is an important element of local identity and a link of a community with its history.

Many local traditions are distributed over larger areas and exchange on a regional level reinforces their usage. However, many of these local traditions escaped the attention of a broader public in these regions or the political elite in charge.

The Consortium includes researchers with expertise in a variety of fields allowing a study of such diets from a multifaceted perspective:

- University of London, The School of Pharmacy, Centre for Pharmacognosy and Phytotherapy, London, UK.
- University of Murcia, Department of Plant Biology, Murcia, Spain.
- University of Frankfurt, Biocenter Niederursel, Frankfurt, Germany.
- Jagiellonian University, Institute for Molecular Biology and Biotechnology, Krakow, Poland.
- University of Milan, Department of Pharmacological Sciences, Milan, Italy.
- Harokopio University, Department of Nutrition and Dietetics, Athens, Greece.
- DSM Nutritional Products, Basel, Switzerland.

The Consortium research will hopefully contribute to the development of new nutraceuticals by identifying plants traditionally used in rural communities of southern Italy, Greece and southern Spain with potential health beneficial effects, but as importantly the project wants to give new values to local food products which have been used for many generations and which now are at the brink of becoming forgotten (see Introduction: The Changing Food Pattern).

Specific goals of the project include:

(1) Ethnobotanical documentation of food products of selected communities in southern Italy, Spain, Greece, incl. Crete.
(2) Comprehensive understanding of the social, cultural, economical framework of local food use.
(3) Identification of active extracts/pure compounds (leads for new health food supplements).
(4) Biochemical/pharmacological in vitro mechanisms/in vivo effects of selected species.
(5) Dissemination of ethnobotanical information in local/national languages.

Before discussing this approach in more detail, it is essential to discuss the responsibilities of the researchers and their institutions involved in such projects.

Benefit Sharing

As is apparent from the above, there are two distinct but connected lines of research activities in this project – the study of local food knowledge with the goal of making this of interest to future generations in the regions of origins and thus to contribute to sustainable local use of resources, and the study of the species' wider potential as nutraceutical/health foods. These two lines are in no way based on contradictory goals, but in fact interdependent aspects of one common goal – *the valorization of local knowledge*. Importantly, valorization has two aspects – an economic and a cultural one. Such knowledge will only be 'preserved for posteriority' if it is of interest and value to the community and if it has the potential to contribute to local agriculture. Clearly the wider use of one or a few of these food products would be an opportunity for such a valorization.

In order to ascertain appropriate benefit sharing between the regions of origin and the academic institutions involved in the project, the Consortium has given itself a precise framework in the form of a Consortium agreement. The overall vision is best summarized in the preamble of this agreement, which is legally binding for all partners of the project:

'It is the goal of this research project and the joint interest of all partners to evaluate food plants locally used in member countries of the EU and neighboring countries in nutrition which have recorded additional health benefits; to characterize these plants with respect to their biological and pharmacological effects; to study the pharmacological mechanism of action of selected plant extracts in order to better understand the biochemical mechanism and thus to evaluate these popularly used local food plants; to contribute to the dissemination of knowledge about locally used resources in the member countries of the EU with the long-term goal of providing additional incomes to the regions of origin.

It is the explicit principle that the botanical material provided by the partners and the various external partners (subcontractors and any other providers) should be provided under the principles as they are laid out in the 'Convention of Rio' and subsequent international agreements. Specifically, plant samples will only be included in the research project if the provider presents documentary evidence that they have the permission from appropriate government authorities to collect these plants.

It is understood by all parties signing this agreement, that information on the bioactivity and chemical characterization of collected species will not be used in conflict with the sustainable use of these resources and that endangered and rare species are not to be overexploited, regardless of their potential as patentable pharmaceutical agents with commercial value.

Specifically, mechanisms have to be set up for just compensation of indigenous groups or other traditional or local keepers of knowledge, if the plant samples provided by them yield economical benefits. It is recognized that this compensation refers to all commercial products such as nutraceuticals or pharmaceuticals, whose discovery and development is based, either directly or indirectly, on the TK of the respective groups.'

The most crucial aspect clearly relates to the potential of developing a patentable product, which could generate significant income for one or several of the partners involved. In such a case the Consortium 'Local Food-Nutraceuticals' will set up a Trust Fund, which will be in charge of the equitable sharing of the financial benefits of the project: 'In the event of payment to the Trust Fund a Benefit Sharing Fund will make donations to return to the country or region which supplied the botanical material to support conservation, health training and education at the community level or regional level when the source material originates from an uninhabited area' (Consortium Agreement 'Local Food-Nutraceuticals', § 10.5.5). Of the total revenues, 40% will be returned to the regions of origin.

Currently (June 2006) no such product has been developed, but clearly the agreements remain legally binding for the future activities of the partners on the species included in this project.

The Consortium agreement has been phrased in a way to allow for its use within the EU or beyond. We selected this model since the likelihood of developing promising new lead extracts or compounds is certainly higher than if one would have used a random approach, but it is still rather low. Therefore, such measures will only be implemented if needed. It has often been argued that the communities in which such research is conducted should be reimbursed for providing the knowledge as such without waiting for potential but highly uncertain future benefits. However, such funds were not made available under the fifth framework programme of the EU and on an EU-wide level the relevant policy issues on this local technological knowledge have not been resolved.

The Nutritional-Pharmacological Angle

As part of the project the biological activities of the extracts were assessed in twelve different assays covering a broad range of mechanisms considered crucial in the pathology of chronic, aging-related diseases [for details, see 35]. In simple terms, a certain number of extracts are taken through a series of in vitro tests with the ultimate goal to identify species which are of particular potential for further development (fig. 1). Four antioxidant tests: DPPH scavenging, prevention of oxyhemoglobin bleaching, prevention of lipid peroxidation (malondialdehyde formation), and protection from DNA damage (Comet assay); three enzyme inhibition tests: inhibition of xanthine oxidase, inhibition of myeloperoxidase-catalyzed guaiacol oxidation as well as the inhibition of

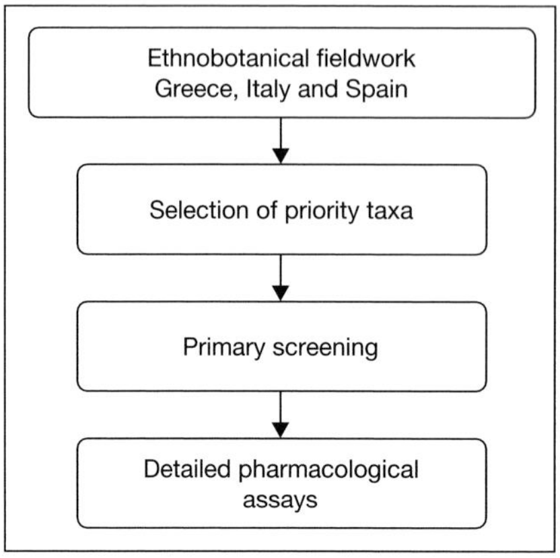

Fig. 1. Overall strategy of the pharmacological-nutritional angle of the Consortium's research.

acetylcholine esterase; one test investigating the inhibition of cytokine-induced cell activation (including the extracts' potential cytotoxicity); one assay measuring the antiproliferation potential; one test assessing the antidiabetic activity (PPARγ) as well as one assay investigating the extracts' effect on mood disorder-related biochemical parameters (hSERT). Furthermore, the polyphenol content of all extracts was determined using the Folin-Ciocalteaus method. The assays revealed diverse biological effects for the tested extracts ranging from no activity to almost complete inhibition/activation. Moreover, the experimental matrix led to the identification of a subset of extracts – i.e. *Berberis vulgaris, Reichardia picroides, Scandix australis, Satureja montana, Thymus piperella, Lythrum salicaria*, and *Vitis vinifera* – showing high activity in a broad number of assays. In a nutshell, the modulations observed in vitro and effects exerted by extracts derived from local food plants suggest that these plants may contribute to the observed better aging of rural Mediterranean populations [35–37].

Local Dissemination of Knowledge

On important outcome of this project has been the public dissemination of such knowledge. The project is not just one of academic scholarship or the industrial development of novel products but aims at transmitting this knowledge to future generations. Therefore a number of activities form part of the project. The most important ones are the following:

- *In Italian*: 'Ta chòrta – wild food plants of Gallicianò' [30] summarizing the information for the Greek-Italian community of Gallicianò and is, among others, made available to members of the Magna-Graecian communities.

- *In Spanish*: Development of the Curricular Materials 'La alimentación en Castilla La Mancha. De la escasez al desperdicio (el valor de los alimentos locales y su utilización sostenible)'. The materials received an award from the Regional Education Authorities in 2004, and were published in digital form [38], including open-source interactive puzzles with images of local food types. These materials are available in all secondary education centers in Castilla La Mancha (>400) with a potential public audience (school children) of >100,000.
- *In Spanish*: A poster exhibition explaining basic data on GFPs and fungi. This exhibition has been on loan to different local organisms for display since November 2004. Over 20,000 people visited this exhibition at the Food and Agriculture Fair in Albacete in December 2004.
- *In Spanish*: A book addressed to the general public in Spain: 'Guía etnobotánica de los alimentos locales recolectados en la provincia de Albacete' [27]. In this book local uses of approximately 150 GFPs and fungi from the provinces of Albacete and Cuenca are recorded and explained (with a repertory of 200 traditional recipes). It is richly illustrated, organized as a guide, with color photographs of the plants, the parts used and of traditional dishes. This book includes a brief summary of the most salient results prepared by the research groups involved in pharmacological research on the selected Spanish species.

These examples highlight approaches to disseminate such knowledge on a local level, but numerous other ones ought to be explored, like local food and medicinal plant gardens, didactic material for primary schools and collaborations with local national parks or museums. Several examples of the development of such locally useful outputs of projects have been published over the last decades and it also will be essential to get the support from international (like the EU-Framework programmes) and national funding agencies. So far this has largely been a domain of based on individual or non-governmental initiatives with funding by local authorities.

Conclusions

In this review we discussed theoretical concepts, which are crucial to understand the role of gathered and locally cultivated food plants (including fungi) in communities of the Mediterranean. These concepts form the basis of a multidisciplinary research project which is summarized in several of the following chapters. The idea that regional and local cultivars or dishes are an important element of many cultures is, of course, not a new one. However, the *concept of 'local food' now offers a theoretical framework, which allows for the study and further development of these crucial elements of 'local knowledge'.*

This framework resulted in one specific project, which is also summarized, pointing to the potential of such a multidisciplinary approach. Specific outcomes and limitations are discussed and reviewed in many of the subsequent chapters of this volume.

References

1 Donaldson MS: Nutrition and cancer: a review of the evidence for an anti-cancer diet. Nutr J 2004;3:19. doi:10.1186/1475-2891-3-19 (http://www.nutritionj.com/content/3/1/19).

2 Parikh P, McDaniel MC, Ashen MD, Miller JI, Sorrentino M, Chan V, Blumenthal RS, Sperling LS: Diets and cardiovascular disease: an evidence-based assessment. J Am Coll Cardiol 2005;45: 1379–1387.

3 Fujimoto WF: Overview of non-insulin-dependent diabetes mellitus (NIDDM) in different population groups. Diabet Med 1996;13(suppl 6):S7–S10.

4 Shaw JE, Chrisholm DJ: Epidemiology and prevention of type 2 diabetes and the metabolic syndrome. Med J Aust 2003;179:379–383.

5 Narayan KM, Kanaya AM, Gregg EW: Lifestyle intervention for the prevention of type 2 diabetes mellitus: putting theory to practice. Treat Endocrinol 2003;2:315–320.

6 Rivera JA, Barquera S, Gonzalez-Cossio T, Olaiz G, Sepulveda J: Nutrition transition in Mexico and in other Latin American countries. Nutr Rev 2004;62:S149–S157.

7 Andrade-Cetto A, Heinrich M: Mexican plants with hypoglycaemic effect used in the treatment of diabetes. J Ethnopharmacol 2005;99:325–348.

8 Ramachandran A: Epidemiology of diabetes in India – three decades of research. J Assoc Physicians India 2005;53:34–38.

9 World Health Organization: http://www.who.org (accessed 06/07/2005).

10 Keys A, Keys M: How to Eat Well and Stay Well: The Mediterranean Way. New York, Doubledays, 1959.

11 Hoffman W: Meet Monsieur Cholesterol. Minneapolis, University of Minnesota, 1979. http://mbbnet.umn.edu/hoff/hoff_ak.html (accessed 06/07/2005).

12 Keys A: Seven countries: A Multivariate Analysis of Death and Coronary Heart Disease. Cambridge, Harvard University Press, 1980.

13 Wiersum KF: Incorporating indigenous knowledge in formal forest management: adaptation or paradigm change in tropical forestry? In Lawrence A (ed): Forestry, Forest Users and Research: New Ways of Learning. Wageningen, ETFRN, 2000, Series No 1, pp 19–32.

14 Acre A, Long N: The dynamics of knowledge-interfaces between bureaucrats and peasants; in Long N, Long A (eds): Battlefields of Knowledge: The Interlocking of Theory and Practice in Social Research and Development. London, Routeledge, 1992, pp 211–247.

15 Lévi-Strauss C: L'origine des manières de table. Mythologiques 3. Paris, Plon, 1968.

16 Geertz C: Local Knowledge – Further Essays in Interpretative Anthropology. New York, Basic Books, 1983.

17 Johnson M (ed): LORE: Capturing Traditional Environmental Knowledge. Ottawa, Dene Cultural Institute and IDRC, 1992.

18 Berkes F, Colding J, Folke C: Rediscovery of traditional ecological knowledge as adaptive management. Ecol Appl 2000;10:1251–1262.

19 Gadgil M, Berkes F, Folke C: Indigenous knowledge for biodiversity conservation. Ambio 1993;22: 151–156.

20 Berkes F: Sacred Ecology. Traditional Ecological Knowledge and Resource Management. Philadelphia, Taylor & Francis, 1999.

21 Secretariat to the Convention on Biological Diversity: Handbook of the Convention on Biological Diversity. London, Earthscan Publications, 2001.
22 Anderson EN: Ecologies of the Heart: Emotion, Belief, and the Environment. New York, Oxford University Press, 1996.
23 Pierotti R, Wildcat D: Traditional ecological knowledge: the third alternative (commentary). Ecol Appl 2000;10:1333–1340.
24 Askegaard S, Madsen TK: The local and the global: exploring traits of homogeneity and heterogeneity in European food cultures. Int Bus Rev 1998;7:549–568.
25 Fieldhouse P: Food and Nutrition: Customs and Culture. London, Croom Helm, 1986.
26 De Cervantes M: Don Quixote. http://www.readprint.com/work-185/Miguel-de-Cervantes (accessed 5/7/2005).
27 Rivera D, Verde A, Fajardo J, Inocencio C, Obón C, Heinrich M (eds): Guía Etnobotánica de los Alimentos Locales Recolectados en la Provincia de Albacete. Albacete, Instituto de Estudios Albacetenses 'Don Juan Manuel', 2005.
28 Rivera D, Obón C, Inocencio C, Heinrich M, Verde A, Fajardo J, Llorach R: The ethnobotanical study of local Mediterranean food plants as medicinal resources in southern Spain. J Pharmacol Physiol 2005;56(suppl):97–114.
29 Mennell S, Murcott A, van Otterloo A: The Sociology of Food: Eating, Diet, and Culture. London, Sage, 1992.
30 Nebel S: Ta chòrta: Piante commestibili tradizionali di Gallicianò. London, University of London, School of Pharmacy, 2005.
31 Hocking GM: A Dictionary of Natural Products. Medford/NJ, Plexus Publications, 1997.
32 Pieroni A, Nebel S, Santoro RF, Heinrich M: Food for two seasons: culinary uses of non-cultivated local vegetables and mushrooms in a South Italian village. Int J Food Sci Nutr 2005;56:245–272.
33 Heinrich M, Leonti M, Nebel S, Peschel W: 'Local Food-Nutraceuticals': an example of a multidisciplinary research project on local knowledge. J Physiol Pharmacol 2005;56:5–22.
34 Inocencio C, Rivera D, Obón C, Verde A, Heinrich M: Local Food as a Sustainable Development Issue in SE Spain. ISE Congress, Pretoria 2002.
35 The Local Food-Nutraceutical Consortium: Understanding local Mediterranean diets: a multidisciplinary pharmacological and ethnobotanical approach. Pharmacol Res 2005;52:353–366.
36 Grande S, Bogani P, De Saizieu A, Schueler G, Galli C, Visioli F: Vasomodulating potential of Mediterranean wild plant extracts. J Agric Food Chem 2004;52:5021–5026.
37 Schaffer S, Eckert GP, Müller WE, Llorach R, Rivera D, Grande S, Galli C, Visioli F: Hypochlorous acid scavenging properties of local Mediterranean plant foods. Lipids 2004;39: 1239–1247.
38 Verde A, Fajardo J, Rivera D, Obón C, Inocencio C, Heinrich M: Materiales Curriculares 'La alimentación en Castilla-La Mancha: de la escasez al desperdicio. (El valor de los alimentos locales y su utilización sostenible)'. Murcia, Azarbe, 2004, pp 1–175 (ISBN 84–96299–23–6, Digital).

Michael Heinrich
Centre for Pharmacognosy and Phytotherapy
The School of Pharmacy, University of London
29–39 Brunswick Square, London WC1N 1AX (UK)
Tel. +44 20 7753 5844, Fax +44 20 7753 5909, E-Mail phyto@pharmacy.ac.uk

Heinrich M, Müller WE, Galli C (eds): Local Mediterranean Food Plants and Nutraceuticals. Forum Nutr. Basel, Karger, 2006, vol 59, pp 18–74

Gathered Mediterranean Food Plants – Ethnobotanical Investigations and Historical Development

Diego Rivera[a], *Concepción Obón*[b], *Michael Heinrich*[c], *Cristina Inocencio*[a], *Alonso Verde*[a], *José Fajardo*[a]

[a]Departamento de Biología Vegetal, Facultad de Biología, Universidad de Murcia, Murcia, [b]Departamento de Biología Aplicada, EPSO, Universidad Miguel Hernández, Orihuela, Alicante, Spain; [c]Centre for Pharmacognosy and Phytotherapy, The School of Pharmacy, University of London, London, UK

Abstract

The diversity of local Mediterranean food elements is not known in detail, but offers itself to search for new vegetables, salads, fruits and spices which could be used in to enrich diets outside their region of origin. Most amid those interesting local elements are edible wild plants and weeds. Ethnobotanical research has identified ca. 2,300 different plant and fungi taxa, which are gathered and consumed in the Mediterranean. Among these, >1,000 are only consumed in one single zone, therefore are strictly local. The percentage of local gathered food plant (GFP) taxa (present in <5 samples), is higher in the main centers of diversity at the periphery of the Mediterranean (Sahara, Alps, Caucasus, Canary Islands, the Levant). Islands (Sicily, Sardinia, Crete, Cyprus) also show a high proportion. Endemism of GFP taxa only accounts for a limited number of these 'ethnobotanical endemics' (only ca. 350 are endemic/endangered species). On the other hand, only a few taxa – 30 occurring in >20 samples – are consumed in most of the Mediterranean. Most have been analyzed in the Local Food-Nutraceuticals project. The ca. 800 GFP taxa that occur in more than the 5% of localities show a geographical pattern that permits one to recognize seven geographical groups. These groups show relationships with types of Mediterranean diet and could also be related with human genetic polymorphism through long-term co-evolution in a geographical mosaic pattern.

For many people the idea of a Mediterranean diet suggests the combination of a diversity of vegetables, salads, fruits and spices, which are often derived from local traditions, including some widely known basic ingredients (pasta, olive oil, wine). However, the diversity of local food elements is in fact not known in detail, but offers itself to search for new vegetables, salads, fruits and spices which could be used in to enrich diets outside their region of origin. Most among those interesting local elements are edible wild plants and weeds.

There are several exhaustive global catalogues of edible plants of the world, which include crops, wild plants and weeds [1–3]. Similar information is available for individual countries around the Mediterranean [4–6]. Local and regional floras have been published covering most Mediterranean countries. Therefore by collating both sources we can easily obtain a list of local wild edible plants. Then what makes the difference of the ethnobotanical study of local food in the Mediterranean area? Usually we discover a smaller repertory of taxa. People do not eat all 'edible' plants available in their environments but only a small part of this flora. What makes the difference is the cultural decision that is behind each repertory of gathered food plants (GFPs).

Ethnobotany is the interdisciplinary study of plant-human relationships embedded in a complex and dynamic system of natural and social components [7, 8]. Ethnobotanical approaches to the study of Mediterranean food plants offers novel ways for analyzing and preserving traditional knowledge and agrobiodiversity in the Mediterranean area. However, these food plants do no stand in isolation, but in the complex interplay of plant diversity, human cultural and genetic diversity.

Ethnobotany can help to determine precisely which plants are currently consumed by each ethnic group in a determined geographical and cultural context. However, why is this relevant? The answer is that in simple terms an edible plant is not necessarily eaten. There are many different factors that determine the choice of a specific species as a food: abundance, availability, cultural preferences, processing technologies, ability to collect it in the optimal period and, last but not least, genetic features of the consumers (e.g. presence of detoxifying enzymes) that allow the safe consumption of the plant. Therefore, most 'edible' plants are not actually consumed in localities where they are locally abundant. Sometimes they are only used as fodder, or included in the restrictive group of 'famine food', or simply neglected. Ethnobotany shows that this selective local profile of food is found at different scales (local to regional) and of course it is part of individual traditional knowledge systems (TKS).

While cultivated plants come to mind first as elements of food plant diversity, GFPs are crucial for understanding the health impact of these diets. Their contribution to the traditional Mediterranean diet, although qualitatively important, is yet almost unknown and offer a unique opportunity for discovering and developing potential new products and thus crops [9–11].

Consequently, GFPs not only offer unique opportunities to develop new products and crops, but even more importantly the topic requires a thorough diachronic scientific analysis and discussion from a multidisciplinary perspective:

- What species are still available and in which regions of the Mediterranean?

- What is the current contribution of GFPs to Mediterranean diets at a local level?
- What are the mechanisms that allow the conservation of and innovation in local repertories of GFPs?
- How these are integrated into traditional local recipes and combined with other ingredients?
- How do factors like season of gathering, initial processing, cooking and combination in complex recipes influence the palatability, nutritional value and health benefits of GFPs?
- How likely is it that much of this knowledge and diversity will to be transmitted to future generations, and, hence, how much is likely to be lost?

Here we propose that an ethnobotanical framework offers some crucial answers to these and related questions. This theoretical framework is, in part, based on human cultural and genetic diversity and their interactions and before discussing the interplay between the various factors, we summarize the relevant data briefly.

Cavalli-Sforza and Feldman [12] suggest that co-evolution of genes with language and some slowly evolving cultural traits supports and supplements the standard model of human genetic evolution. Advances in our understanding of the evolutionary history of humans attest to the advantages of multidisciplinary research. However, the role of plants in human evolution needs to be evaluated as a co-evolutionary process. Co-evolution supposes mutual relationships – tolerance, resistance and fitness – between species: predator/prey, herbivore/plant consumed. Human consumption of plant food is a particular case of herbivory. It opens a wide range of possibilities in terms of adaptive changes, both in plant and human populations. Changes in plant species (genetic, morphological, physiological) are associated to 'domestication'. These are apparent if one compares cultivated plants with their wild relatives. Human selection acted in favor of tolerance, viz. plants able to furnish a substantial crop (parts suitable for human consumption) without negative fitness consequences. The intentional or unintentional modification of their habitats led in different places and moments to the origins of agriculture and forestry in form of geographical mosaics. The geographical-mosaic pattern of co-evolutionary interactions appears to be ubiquitous in nature [13]. Geographical mosaics are the output of human activities either as hunter-gatherers or as agriculturalists, they are typical of the Mediterranean landscapes.

Much less explored aspects are the adaptive changes in human populations as a consequence of continued consumption of a particular type of food. It is evident that food offers to the human beings the possibility of interacting with a wide range of substances (not only nutrients). The interaction between these ingested compounds and the expression of genes is extremely relevant in terms of the difference in fitness between the damaged and undamaged subjects. This led to a differential concept of food (or a less safe food) in function of the genetic profile of the food consumer. Diet may be the most influential environmental factor modulating the phenotypes for both monogenic and multifactorial diseases.

There are many examples of genetic inadaptation of humans to nutrients and other compounds in food. Some well-known ones are: phenylketonuria, clinical galactosemia, lactose intolerance, celiac disease, familial hypercholesterolemia, congenital hyperhomocystinemia, cystinuria, hyperuricemia, and hemochromatosis. The prevalence of these inadaptation syndromes shows significant differences among ethnic groups and their fitness impact is mediated through diet. However, some are seen on the contrary not as 'inadaptation' but as adaptive success in ethnic groups long exposed to the substance. The genetic trait in which intestinal lactase activity persists at childhood levels into adulthood, varies in frequency in different human populations, being most frequent in Northern Europeans (>85%) and certain African and Arabic nomadic tribes, who have a history of drinking fresh milk. Selection is likely to have played an important role in establishing these different frequencies since the development of agricultural pastoralism ca. 9,000 years ago [14].

Celiac disease results from a permanent intolerance to ingested gluten and related cereal prolamins in genetically predisposed individuals. The deleterious proteins are gliadins (wheat), hordeins (barley), secalins (rye), and possibly avidins (oats) [15]. As a heritable condition, familial aggregation is common, and it is more prevalent in Caucasians. Have Caucasians supported a relative low exposure to negative selection by rich gluten food in their history?

Adaptive features are also found in cultural patterns (as part of the TKS), e.g. the detoxification practices (boiling, simmering, fermentation) for plant food rich in alkaloids, steroids, sequiterpene lactones, etc. Also, precise distinction of the timing for collection helps to avoid toxic substances in excess; often this is associated with dates of the religious calendar and festivals.

The integration of genetic and cultural diversity in terms of food and food patterns is still a major endeavor. Ethnobotany is in many regions of the world by necessity a diachronic discipline, but it needs to become more synchronic and contribute to an understanding of the role of plants in human history and even evolution [16, 17]. With this review we wish to contribute to understand the deep meaning of the saying 'You are what you eat'.

Human Genetic Diversity

The first aspect we need to focus on is the human genetic diversity of the Mediterranean characterized by successive colonization episodes. These involve the presence of scattered glacial refugia within the area. Colonization occurred not only westwards (Paleolithic before last glacial maximum (LGM), Neolithic)

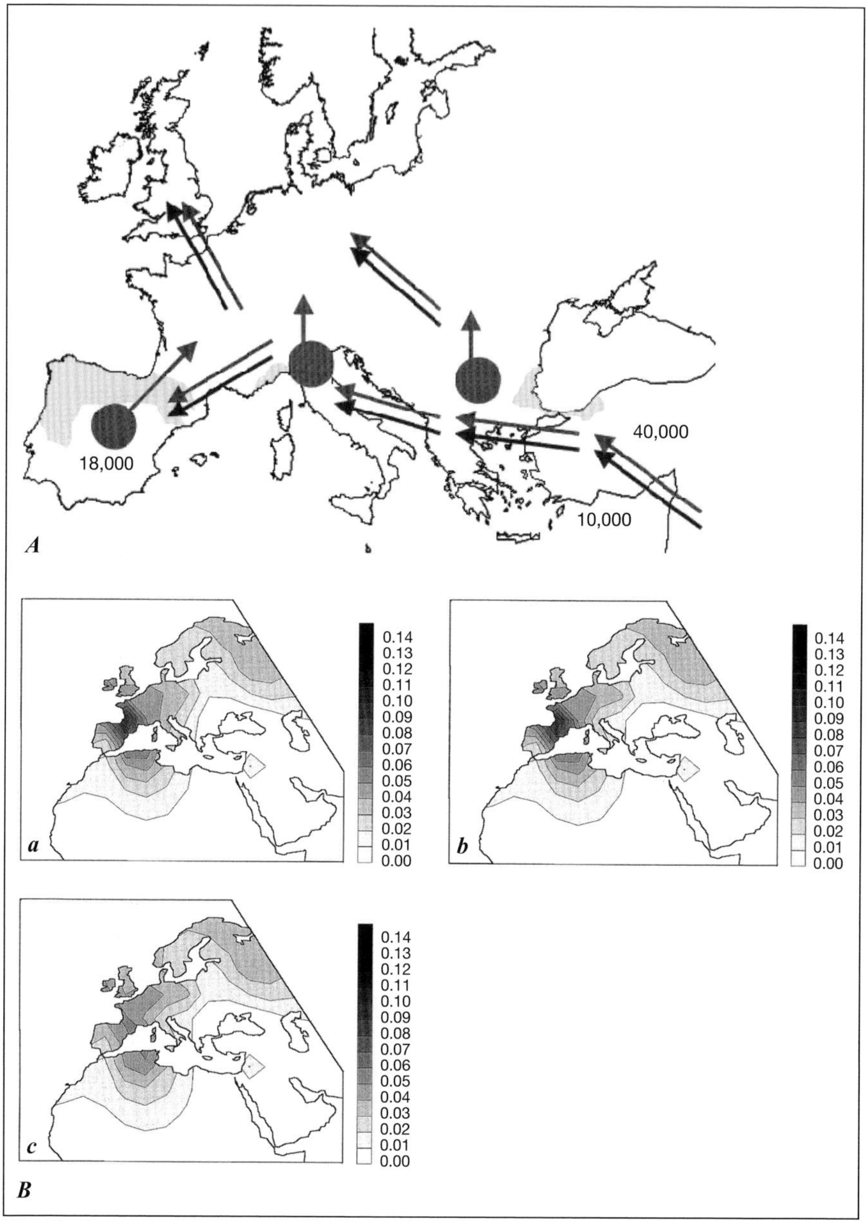

Fig. 1. *A* A scheme of the main demographic processes documented in the archaeological record of Europe. Numbers are approximate dates, in years before the present. Left-pointed arrows, above, Palaeolithic colonization; up-pointed arrows, Mesolithic reexpansions (glacial refugia are represented by circles); left-pointed arrows, below, Neolithic demic diffusion [38]. The suggested homeland for mitochondrial DNA haplogroup V is the population refugium in

but also eastwards (Paleolithic after LGM, post-Neolithic) (fig. 1, table 1), giving place to interactions with the different local floras and, most likely, to pre-agricultural plant introductions both in the west and in the east of the Mediterranean. The time and place of these episodes is best determined by analyzing human genetic polymorphism. The relative contribution of the different colonization episodes to the present human genetic polymorphism is highly controversial, especially those of the Pre-Neolithic hunter-gatherers: estimates range from <30% [18] to >80% [19, 20] (table 1).

Neanderthals left no genetic imprint on modern humans and there is no evidence of any Neanderthal/Modern human hybrids [21, 22]. These statements have been challenged with (still unclear) hybrids of Portuguese Lagar Velho and from Central Europe [23, 24].

The events leading to the emergence of modern humans in Middle Pleistocene were not restricted to one region of the world alone. The last 50,000–150,000 years of human history have been characterized by rapid demographic expansions and the colonization of novel environments outside of sub-Saharan Africa [24]. Genetic adaptation to local environmental conditions may have been more prevalent in colonizing populations [25, using genome-wide patterns of DNA polymorphism, 26]. The gradual loss of diversity in successive colonization bottlenecks (periods of small population size) as our species grew and spread from Africa is suggested by microsatellites showing a clear, almost linear diversity gradient with a high rate in Africa and a decline outside [27].

Two major subsets of mitochondrial deoxyribonucleic acids (mtDNAs) were identified in populations of Western Eurasia and Northern Africa. One is haplogroup V proper, and the other has been termed 'pre V,' since it predates V phylogenetically. The rather uncommon pre V tends to be scattered throughout Europe (and northwestern Africa), whereas V attains two peaks of frequency: one situated in southwestern Europe and one in the Saami of Northern Scandinavia. Geographical distributions (conspicuous, though scattered, occurrence of potential pre V clades in the Mediterranean) and ages support that pre V originated (like its larger sister haplogroup H), before the LGM (21,000– 28,000 BP), perhaps in Eastern Europe, from where it spread along an east-west axis with gravettian contacts (28,000–21,000 BP) [28]. The haplogroup V arose in a southwestern European refugium soon after the LGM. The distribution of haplogroup V mtDNAs in modern European populations would thus, at least in part, reflect the pattern of postglacial human recolonization from that refugium. Except for rare, isolated

Iberia during the Last Glacial Maximum (LGM). It would be diffused with the dispersal of population over the period from 18,000 to 10,000 BP [22, 28, 141]. ***B*** Spatial frequency distributions of V mtDNAs, excluding possible outlier populations: (***a***) without Saami; (***b***) without Saami and Croatians, and (***c***) without Saami, Croatians, and Basques [28].

Table 1a. Main human co-evolutionary groups in function of linguistic, genetic, diet type and GFP type in the Mediterranean

Type	Group 1	Group 2	Group 3	Group 4	Group 5	Group 6	Group 7	Group 8	Group 9	Group 10	Group 11	Group 12
Language	Western Indo-European (Romance)	Eastern Indo-European (Greek)	Eastern Indo-European (Slavic, Albanian)	Eastern Indo-European (Armenian)	Western Afro-Asiatic (Berber)	Western Afro-Asiatic (Berber)	Eastern Afro-Asiatic or Semitic (Arabic, Hebrew, Maltese)	Eastern Afro-Asiatic or Semitic (Arabic, Hebrew)	Basque	Altaic (Turkish)	Altaic (Azeri)	South and North Caucasian
Main Genetic component	Paleolithic	Neolithic	Neolithic	Paleolithic?	Neolithic Near Eastern	Neolithic Near Eastern	Neolithic Near Eastern	Neolithic Near Eastern	Paleolithic	Post-Neolithic?	Post-Neolithic?	Paleolithic
GFP type	Western Mediterranean	Eastern Mediterranean	Alpine	Alpine (Caucasian)	North African	Saharan	North African	Levantine	Western Mediterranean	Eastern Mediterranean	Alpine (Caucasian)	Alpine (Caucasian)
Diet type	Western Mediterranean	Eastern Mediterranean	Adriatic	Caucasian	North African	North African?	North African	Eastern Mediterranean	Western Mediterranean?	Eastern Mediterranean	Caucasian	Caucasian

For explanation of main genetic components see table 1b, for GFP types see table 2a, and for diet types see table 6.

Table 1b. Present human genetic polymorphism (mitochondrial DNA) represents a summary of the three main waves of European colonization

Component	Dates (BP)	Main associated clusters mtDNA	Contribution to modern gene pool, %		Direction of wave
			A*	B*	
Neanderthal	300,000	Unclassified	0	0	E to W?
Early Upper Paleolithic	40,000–50,000	U5	10	10	E to W
Late Upper Paleolithic	11,000–18,000	H, V, I, W, T, K	70	20	W to N and E
Neolithic	8,500–10,000	J (+ more of H, T, K?)	20	70	E to W

The letters refer to the classification of haplotypes. It should be noted that the Neolithic contribution is estimated at ca. 20% of the total by most authors.
*Sources: A: 19, 20, 22, 38, 139; B: 18, 140.

occurrences, V is virtually absent in the Southern Balkans, Turkey, the Caucasus, and the Near East. With regard to age and frequency, there is a clear cline from west to east. The estimated age for V in the west (ca. 16,000 years) is almost twice that in the east, indicating the direction of settlement (fig. 1). The older age reflects the onset of the recolonization of Europe from western refugia. In contrast, the distribution of pre V suggests that these rather rare mtDNAs must have been present in more than one refuge area [28]. Studies in haplogroup H support in part the expansion from the Iberian Peninsula (Franco-Cantabrian Glacial Refuge). However, estimated age for subhaplogroups and relative frequencies are somewhat contradictory [29–31]. Figure 1B also highlights the simultaneous recolonization of North Africa from the eastern littoral of Algeria. This was presumably occurring after the end of the LGM, associated to the subsequent drastic reduction of the Sahara desert extension during the Holocene Climatic Optimum. Analyses of mtDNA sequence variation in European populations suggest that the gene pool has 80% Paleolithic and 20% Neolithic ancestry [20], but still requires the reconstruction and distinction of the gene pools in the various glacial refugia [28].

Richards et al. [32], applying the founder analysis of non-recombining mtDNA sequence data, conclude that (i) there has been substantial back-migration into the Near East; (ii) the majority of extant mtDNA lineages (haplogroups H, pre V, and U5) either originated or have spread along the Mediterranean into Europe in several waves during the Upper Paleolithic; (iii) there was a founder effect or bottleneck associated with the LGM (21,000 BP), from which derives the largest fraction of surviving lineages, and (iv) the immigrant Neolithic component (haplogroups J, T1, and U3) is likely to comprise less than one-quarter of

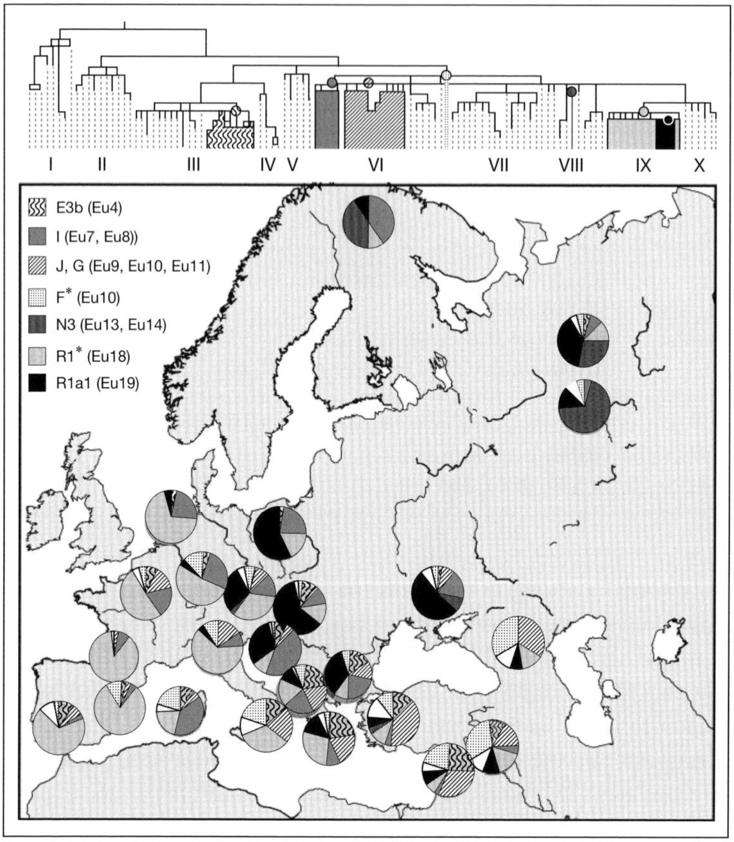

Fig. 2. Maximum parsimony tree for the major Y-chromosome haplogroups and their distribution in Europe [19]. The West Eurasian parts of the tree are highlighted and gray-scale/-pattern-coded. The haplogroup nomenclature of Underhill et al. [142] is indicated beneath the tree; that of the Y-Chromosome Consortium [264] is within the frame, and that of Semino et al. [20] is in parentheses. An asterisk indicates a potentially paraphyletic lineage [264]. See Semino et al. [20] and Richards [19] for further details.

the mtDNA pool of modern Europeans. The haplogroup (pre-HV)1, by contrast, did spread along the Mediterranean either during the Neolithic period or in more recent times or both [33].

Geographic distribution and age estimates of alleles in non-recombining Y-chromosome are compatible with two Paleolithic and one Neolithic migratory episode that have contributed to the modern European gene pool. Scozzari et al. [34] analyzed Y-chromosome variation in 1,127 males from the Western Mediterranean area, northeastern Europe and the Middle East. The present distribution of one of the 17 haplogroups identified suggests that the Neolithic gene pool had a major impact in the eastern and central part of the Mediterranean basin, but very limited consequences in Iberia and northwestern Europe [35, 36] (fig. 2). Two other haplogroups were found to have much more restricted geographic distributions. One of them is common only in Sardinia, which confirms the genetic peculiarity and isolation of the Sardinians.

Using the allelic variability at six Y-chromosome-specific polymorphisms in 333 males from 12 populations around the Mediterranean Sea a 'core' of populations directly facing the north and the east of the Mediterranean basin, from the Middle East to the Italian Peninsula, was found to be genetically undifferentiated. Through the analysis of microsatellite variation, the time to the most recent common ancestor has been estimated to be quite recent (7,000–11,000 BP) [37].

Some differences exist between the mtDNA and Y data pertaining to the putative Paleolithic components. These discrepancies may be due in part to the apparent more recent molecular age of Y-chromosomes relative to other loci. Gender-based differential migratory demographic behaviors will also influence the observed patterns of mtDNA and Y variation [20]. Both mtDNA and Y-chromosome analyses have indicated a contribution of Neolithic Near Eastern lineages to the gene pool of modern Europeans of around a quarter or less. This suggests that dispersals bringing the Neolithic to Europe may have been demographically minor and that contact and assimilation had an important role [19].

In European populations specific mutations have a peculiar geographic distribution, which suggests recent local origin of some alleles or input of genes from sources other than the Levant, notably North Africa and Asia [38].

Genetic isolates in the area include islands and continental outliers. All of them share most features of the less isolated Europeans. For instance the Sardinians and Basques, outliers share basically the same Y binary components of the other Europeans. Their peculiar position with respect to frequency is probably a consequence of genetic drift and isolation. Cultural differences may not have led to genetic differentiation [20, 39].

Human Cultural Diversity

Human population structure, inferred from the Y-chromosome, corresponds broadly to language families, but there are examples in which the genetic heterogeneity does not correlate with either linguistic diversity or geographic barriers [40, 41]. A time depth for the origins of language families greater than the generally accepted date of ca. 6,000 years may in some cases be appropriate, allowing for the correlation between language dispersals and demographic processes following the end of the Pleistocene period. The effects of these processes are still visible in the linguistic 'spread zones', considered by Renfrew [22] to be the result of farming dispersals, contrasting with the linguistic 'mosaic zones' whose early origins may sometimes go back to initial colonization episodes during the late Pleistocene period [39]. Nettle and Harris [42] suggest for Eurasia and West Africa that linguistic and genetic phylogenies will only be correlated where there have been large-scale demic diffusions in the last few thousand years, and relative sedentism in the subsequent period.

The native languages around the Mediterranean belong to three major language families: (1) Indo-European (Albanian, Armenian, Greek, Italic, and Slavic subgroups); (2) Afro-Asiatic (Berber, Egyptian, and Semitic subgroups – comprising the latter Arabic and Hebrew), and (3) Altaic (Turkic subgroup). The outlier Basque language is spoken in parts of Spain and France. North Caucasian and South Caucasian language families are present in parts of Russia, Azerbaijan, Georgia, and Turkey in the boundaries of the Mediterranean [43]. In total, 2,200 species have a record as a GFP in the Mediterranean. These are both widespread and local, vascular plants and fungi (table 2). All of them receive names at least in one of the above languages. Tables 3 and 4 present, for selected GFPs, the repertory of names recorded in the different languages. These names offer interesting examples of cognates (words that are related to each other by descent from a common source [cf. 17]). Some of these cognates are found outside the limits of major language families.

There are three theories on the origins of Indo-Europeans: the Indo-European Chalcolithic Invasion, Renfrew's Neolithic Discontinuity and the Paleolithic Continuity [44]. The Marija Gimbutas' *kurgan* theory supports an Indo-European invasion in the Copper Age (4th millennium BC), by horse-riding pastoralists, according to which, the Proto-Indo-Europeans were those who built the Kurgan culture 5th–3rd millennium BC, in the steppe area of Ukraine, north of the Black Sea. From the steppe area, then, in the 3rd millennium BC Indo-European languages were brought to regions all over Europe [45].

Renfrew's theory of Indo-European Neolithic dispersal is based on the observation that the beginning of farming in 7th millennium BC is the only moment in European prehistory that might coincide with the Indo-Europeanization of Europe. Renfrew concluded that these early farmers were the Proto-Indo-Europeans. The two earliest Neolithic cultures of Southern Europe, both dated to the 7th millennium, namely the Balkan complex and the Impresso/Cardial Ware, would represent the first introduction of Indo-European cultures into Europe [46, 47]. However, the two Southern European areas where Neolithic cultures do show infiltrations from the Middle East, are precisely the areas where non-Indo-European linguistic traits are most evident and important: Italy and Greece [48].

The third theory associates the arrival of Indo-European people in Europe and Asia with the colonization of *Homo sapiens* in Europe and Asia from Africa, and not as an event of recent prehistory. The differentiation process of Indo-European

Table 2. Major types of Mediterranean GFP taxa. The 62 zones analyzed (table 5) yielded a total of 2,300 vascular plant and mushroom taxa. The ethnobotanical endemics are 1,050, each used in one single zone. The remaining taxa form 8 groups: 1 widespread and 7 regional

Type	Zone	Characteristic taxa	Notes
Widespread	Used in >33% of the 62 zones, without special geographical pattern	*Portulaca oleracea* L., *Foeniculum vulgare* Mill., *Sonchus oleraceus* L., *Silene vulgaris* (Moench) Garcke, *Cichorium intybus* L., *Rorippa nasturtium aquaticum* (L.) Hayeck, *Asparagus acutifolius* L., *Malva sylvestris* L., *Papaver rhoeas* L., *Rubus ulmifolius* Schott., *Allium ampeloprasum* L., *Arbutus unedo* L., *Crataegus monogyna* Jacq., *Scolymus hispanicus* L., *Borago officinalis* L., *Chondrilla juncea* L.	Weedy species of crop fields. Shrubs and small trees of machia
Alpine	Alps, Pyrenees, Cantabrian range and Caucasus	*Chenopodium bonus henricus* L., *Corylus avellana* L., *Fragaria vesca* L., *Oxalis acetosella* L., *Vaccinium myrtillus* L., *Malva moschata* L., *Polygonum bistorta* L., *Sempervivum* spp., *Carum carvi* L., *Myrrhis odorata* (L.) Scop., *Epilobium angustifolium* L., *Valerianella locusta* (L.) Laterrade, *Artemisia absinthium* L., *Verbena officinalis* L., *Amelanchier ovalis* Medik, *Pastinaca sativa* L., *Abies alba* Mill., *Hippophae rhamnoides* L., *Allium schoenoprasum* L., *Gentiana lutea* L., *Juniperus communis* L.	Wild greens of meadowlands, fresh fruits, nuts, mushrooms and herbs
Caucasian-E Anatolian	Caucasus, Turkey (E)	*Arum elongatum* Steven., *Cotoneaster nummularia* Fisch. & Mey, *Crataegus pontica* C. Koch, *Eremurus spectabilis* Bieb., *Falcaria vulgaris* Bernh., *Malus orientalis* Uglitzk., *Papaver orientale* L., *Rheum ribes* L., *Zosima orientalis* Hoffm.	Herbal cheeses, wild fruits, vegetables
Western Mediterranean	France, Spain, Italy (W)	*Quercus ilex* L., *Malus sylvestris* (L.) Mill., *Crepis vesicaria* L., *Conopodium* spp., *Pinus pinea* L., *Bryonia dioica Jacq.*, *Echium vulgare* L., *Montia fontana* L., *Reichardia picroides* (L.) Roth., *Scorzonera laciniata* L., *Scandix australis* L., *Asparagus albus* L., *Chamaerops humilis* L., *Echium plantagineum* L., *Scorzonera angustifolia* L., *Diplotaxis tenuifolia* (L.) DC., *Cynara humilis* L.	Acorns. Sprouts of plants from the evergreen oak forests and machia

Table 2. (continued)

Type	Zone	Characteristic taxa	Notes
Eastern Mediterranean	Italy (E), Greece, Cyprus, Malta, Turkey (W)	*Asphodeline lutea* (L.) Reichenb., *Eruca sativa* L., *Rubus sanctus* Schreb., *Salvia fruticosa* Mill., *Echinophora tenuifolia* L., *Polygonum cognatum* Meissn., *Muscari comosum* Mill., *Ornithogalum narbonense* L., *Cornus mas* L., *Carum ferulaefolium* Desf., *Tetragonolobus purpureus* Moench., *Sideritis libanotica* Labill.	Bulbs, mustard greens. Plants of the phrygana
Levantine	Turkey (S), Syria, Lebanon, Palestine, Jordan, Israel	*Anethum graveolens* L., *Asphodelus microcarpus* Salzm. & Biv., *Rhus coriaria* L., *Prosopis farcta* (Solander) Macbride, *Gundelia tournefortii* L., *Ficus sycomorus* L.	Herbs, condiments. Cardoons. Shrubs of riparian forests
North African	Morocco, Tunisia, Algeria	*Thymbra capitata* (L.) Cav., *Emex spinosa* (L.) Camp., *Urtica pilulifera* L., *Diplotaxis harra* (Forssk.) Boiss., *Moricandia arvensis* (L.) DC., *Artemisia arborescens* L., *Bunium incrassatum* (Boiss.) Batt., *Atractylis gummifera* L., *Ferula communis* L.	Herbs. Greens of overgrazed areas. Tubers
Saharan	Morocco, Algeria, Libya, Egypt, Saudi Arabia	*Messembryanthemum forskahlei* Hochst., *Rumex vesicarius* L., *Tamarix aphylla* (L.) Karsten., *Alhagi maurorum* DC., *Phoenix dactylifera* L., *Atriplex halimus* L., *Tirmania nivea* (Derf. ex Fr.) Trappe, *Cynomorium coccineum* L., *Asphodelus tenuifolius* Cav.	Small grains, sweet exudates. Underground fungi

Table 3a. Widespread GFP taxa in the Mediterranean and their names in the languages of the Indo-European family [language names after 43]. References: Spanish: [4, 143, 144, 147, 148, 151, 155–158, 160, 161, 163]. Catalan-Valencian-Balear: [111, 145, 149, 150, 152, 154, 159, 162, 164–166]. Provençal and Occitan: [106, 108, 110], Reguis 1878. Corsican: Simonpoli 1993. Sicilian: [121, 122]. Sardinian: [124–126]. Italian: [126, 146, 153, 167, 168]. Arbëreshë: [169]. Albanian: [170]. Greek: [77, 171–173]. Armenian: [5]. Serbian: [172] (cognates and adoptions marked in shades of gray. See also equivalent rows in table 3b)

Language branches	Romance languages (Latin branch)							Hellenic	Armenian (Thracian)	Slavic	Albanian (Illyric)	
Taxon	Spanish	Catalan-Valencian-Balear	Provençal and Occitan	Corsican	Sicilian	Sardinian	Italian	Greek	Armenian	Serbian	Arbëreshë	Albanian
Allium ampeloprasum L.	Ajo porro	All porro	Porri fer	–	Agghiu porru	Porretell	Porraccio	Agriopraso, agrioskordo	–	Praziluk	Qepë salvaç	Qepë veshtash
Arbutus unedo L.	Madroños	Arbosser	Darbousié	Albitru	–	Arboç	Albatra	Kumaria	–	Maginja	–	Mare
Asparagus acutifolius L.	Espárrago triguero	Esparrecs	Asperjo	Saparacu	Sparacogna, sparagnella	Espàrec	Sparaci, asparago selvatico	Agrelli, aghrellia, sparangi	–	Spargla	Sparengjë	Shpargull
Borago officinalis L.	Borrajas	Borratja	Bourragi	Frisgia, burrascia	Vurrania	Moc-moc	Burraggine	Armpeta, Burragtsa	–	Borazina	Vërajnë	Shajë
Chondrilla juncea L.	Achicorias, ajonjera	Morret de bou	Cantalame, lacholebré, sautaovolame	–	Cudidda	Cagliùca	Erba pizzuta, mastrice	Kolla	Koshmoryk		Gjumës	Zegun
Cichorium intybus L.	Achicoria, camarroja	Cama-roja, xicoira	Cicorèia, cicòri, cicoureio fero	–	–	Cama roja, xicòria	Radicchio di campo	Agrioradiko	Ejerdak	Vodopija	Çikour	Çikore
Foeniculum vulgare ssp. *piperitum* (Ucria) Cout.	Hinojo	Fenoll	Fenohl, fenon, fenoun	Finochiu	Finucchieddu rizzu	Fenoll salvatge	Finoccio selvatico	Maratho, marathos	Jorom-samit	Kmorac	Fënoq salvaç	Marajë
Rorippa nasturtium-aquaticum (L.) Hayek	Berros	Créixens	Creisson, creissoun	Cruscione	–	Alxòni	Crescione	Kartamilla, kardhamilles, nerokardamo	Dzhirkotem	Potocarka ugaz	Shërpër	Shelp

Table 3a. (continued)

Language branches	Romance languages (Latin branch)							Hellenic	Armenian (Thracian)	Slavic	Albanian (Illyric)	
Taxon	Spanish	Catalan-Valencian-Balear	Provençal and Occitan	Corsican	Sicilian	Sardinian	Italian	Greek	Armenian	Serbian	Arbëreshë	Albanian
Papaver rhoeas L.	Ababoles, babaoles	Rosella	Rosèla, roela, ruelo	Rosula	Paparina	Papàriu	Pavet, pupattole, rosolaccio	Koutsounada, petinos	Kakatch	Mak	Luljëkuq	Lulkuqe
Portulaca oleracea L.	Verdolaga	Verdalaga, Pulsallana	Bortolaiga, bortoulaigo	Erba fratesca-	Puccidana	Porcellana	Porcacchia	Andraka, glistiria, ghlistridh	Dandyr	Prkos	Burdulak	Burdullak
Scolymus hispanicus L.	Cardillos, tagarninas	Cardo de moro	Pei de nouvé	–	Scoddi, rattameli	Card vaqueta?	Cardo sumarino	Askolimbros, chrysangatho, skolymos	–	Dragusica	Kardunxheljë, rëkoljë	Rrëkuall
Silene vulgaris (Moench) Garcke	Collejas	Conillets, colitxos	Caurilh, cresinèu, petarèla	Scrununietti	Cannatedda	Capriuleddu	Trivoli, stride, strisciola, cucina	Strouthouthkia, tsakrithkia	–	Pusina	–	Klokëz
Sonchus oleraceus L.	Borrajas, cerrajas, cerrajón	Borratxa, llicsó	Engraisso por	–	Cardedda	Card munjòni, Bardu minzone	Allattalepre, Sonco	Çdjokhos, sonchos, trakhouri	Ishamarol	Gorcika	–	Rrëshyell
Urtica dioica L.	Ortigas	Ortiques	Ourtigano, ourtigo	Urticula	Ardica	Occiau, Viltiga	Ortica	Tssouknitha	Ajindsh	Koprina	–	Hithër

Table 3b. Widespread GFP taxa in the Mediterranean in the Afro-Asiatic, Altaic and Caucasian Families of languages. Names are represented not only if the plant is used as food. References: Hebrew: [174–176]. Maltese: [177, 178]. Turkish: [79, 81, 179–183]. Arabic: [6, 84, 85, 115, 116, 184–196]. Berber: [115, 116, 185, 187]. Basque: [197–199, 236]. Azeri: [5]. Georgian: [5] (cognates and adoptions marked in shades of gray. See also equivalent rows in table 3a)

Taxon	Afro-Asiatic				Altaic		Caucasian	Isolate
	Hebrew	Maltese	Arabic	Berber	Turkish	Azeri	Georgian	Basque
Allium ampeloprasum L.	Hazir	Kurrat salvagg	Bussîla, kurrat, tumet el Arab	Kâûl	Kaya sarimsagi, sirmo, sirik	–	–	Basa porrua
Arbutus unedo L.	–	–	Lenj, mothrounia	Metrun, sasnu, sisnou	Kocayemis	–	–	Animania, gurbitza
Asparagus acutifolius L.	–	–	Halyûn	Azzu, tazzût	Yabani kukonmaz	–	–	Basaesparrago
Borago officinalis L.	–	Fidloqqom	Al-hurraycha, lessane-et-tour	Bû-hamdûn	Hodan	–	–	Murriona
Chondrilla juncea L.	–	–	–	–	Karaavlik, Çengel sakizi	Dshingile	Yledjela	Basa-kardabera
Cichorium intybus L.	Merôrîm, les matsvooey	Cikwejra	'Elik, hindaba, seriss	Arhlilou, mersag	Çitlik	Kasni	Bardkatchatcha	Osterchuriya
Foeniculum vulgare ssp. *piperitum* (Ucria) Cout.	Shoomar	Busbies	Besbaça, bisbês, shamra	Ouamsa, tamessaout, semsous	Rezene	Raziana	Dsheredsho	Millua
Papaver rhoeas L.	–	Peprina hamra	Bu gar'ûn, kash-kash, ouadjir	Ftûlû, talûdat	Gülotu, gelincik	Lale	Katcho	Lobelarra

Table 3b. (continued)

Taxon	Afro-Asiatic				Altaic		Caucasian	Isolate
	Hebrew	Maltese	Arabic	Berber	Turkish	Azeri	Georgian	Basque
Portulaca oleracea L.	Rgelat hagenah	Burdlieka	Benderakech, blabicha, redjila	Alora, arrhilem, bouguel, tafrita	Temizlik, tokamkan, parpar	Perpern	Dandyri	Ketozkia
Rorippa nasturtium-aquaticum (L.) Hayek	Gargeeyr hanoochalym, merôrîm	Krexxuni sija	Guernounech	–	Aci tere, suteresi, tere	–	–	Berroa
Scolymus hispanicus L.	Dardar, khokha sfaradee	Xewk isfar	Guernina, quernena, sernish, zirniz	Izifore	Altindikeni, saridiken	–	–	Kardaberaiska
Silene vulgaris (Moench) Garcke	–	Quasqejza	Kahali, nouar ed dil	Talazazt	Givisganotu, tavsan ekmegi, siyavu	–	–	Galkidea
Sonchus oleraceus L.	Meror hageegot	Tfiefa Komuni	Difâf, tifaf	Daouague, wagerrin	Sütlük dikeni	–	Gytcha	Karduntxa
Urtica dioica L.	Kîmmosh	–	Horaig, hurrîqa	Azekdouf, rimezrit	Isirgan, geznil	Gitchitkan	Tchintchari	Asuna, osina

Table 4a. Local GFP taxa in the Mediterranean and their names in the languages of the Indo-European family (language names after Ethnologue 2005). References: Spanish: [4, 143, 144, 147, 148, 151, 155–158, 160, 161, 163]. Catalan-Valencian-Balear: [111, 145, 149, 150, 152, 154, 159, 162, 164–166]. Provençal and Occitan: [106, 108, 110], Reguis 1878. Corsican: Simonpoli 1993. Sicilian: [121, 122]. Sardinian: [124–126]. Italian: [126, 146, 153, 167, 168]. Arbëreshë: [169]. Albanian: [170]. Greek: [77, 171–173]. Armenian: [5]. Serbian: [172] (cognates and adoptions marked in shades of gray. See also equivalent rows in table 4b)

Language branches	Romance languages (Latin branch)							Hellenic	Armenian (Thracian)	Slavic	Albanian (Illyric)	
Taxon	Spanish	Catalan-Valencian-Balear	Provençal and Occitan	Corsican	Sicilian	Sardinian	Italian	Greek	Armenian	Serbian	Arbëreshë	Albanian
Amaranthus graecizans L.	Bledo, bleo dulce	Blets	–	–	–	Amarantu	Amaranto	Amarandon, vliton	Gavakatar		–	Nenë
Anchusa italica Retz. (= *A. azurea* auct.)	Lenguazas	Buglossa	Bourrigas	–	–	Llengua de bou	Buglossa	Budoglossos, oglossos	Gavashynak	Pazje gnezdo	–	Gjuhëlope
Arctium minus Bernh.	Lampazo, gordolobo, ceones	Llepassa borda	Erbo de la jaunisso	–	Guddizzuni	Cuscusone	Lappola	–	Kratyk	Cicak repuh	–	Rrodhe
Beta maritima L.	Acelga de campo, blea	Bleda borda	Bledo	Bietula salvatica	Seccala	Bleda salvatge	Bietola selvatica	Agrioteftlo, agriolahano	–	Blitva	–	Panxhar
Eruca sativa L.	Oruga	Ruqueta	Rouqueto	–	–	Arrucas, Rucca	Rucola	Aromatoss, euzomon to imero, roqqua	–	–	–	–
Eryngium campestre L.	Cardo setero	Panical	Panicau	–	–	Cardu anzoninu	Bocca di ciuco	Aggathia	Eryndjak zrejnak	Vetrovalj	–	Gjembardhë
Lactuca serriola L.	Esperillas, lechuguillas	Lletuguetes bordes	Lachugo fero	–	–	Lletuga salvatge, lattia budra	Erba bussola	Thridax agria	–	Salata	–	Marule, pëfshirë
Lactuca viminea (L.) Presl.	–	–	–		Scursunara	–	Lattuga alata	Agriomaruli	–	–	–	Marule
Malus sylvestris Miller	Maillo	–	–	–	–	Melabrina?	Melo silvatico	Agriomelia	–	Sabuka	Lemonxed	Mollë

Table 4a. (continued)

Language branches	Romance languages (Latin branch)							Hellenic	Armenian (Thracian)	Slavic	Albanian (Illyric)	
Taxon	Spanish	Catalan-Valencian-Balear	Provençal and Occitan	Corsican	Sicilian	Sardinian	Italian	Greek	Armenian	Serbian	Arbëreshë	Albanian
Muscari comosum (L.) Miller = *Leopoldia comosa* (L.) Parl.	–	Caps de moro	Cebouiado	–	Cipudazzu, purrazzu	Mascaró	Lampascione, cipollaccio	Agrioyakinthos, askordoulokas	–	Vilin luk	Çëpuljin	Pështrik cuflor
Myrtus communis L.	Mirto	Murta	Nerto	Morta	–	Murta	Mirto	Myrssini, myrtya	–	Mirta, mrca	–	Mërsinë
Plantago coronopus L.	Rampete	Patetes de pardalet	Bano de cervi	–	–	Erba stella	Barba di cappuccio	Lithospasto	–	Bokvica	–	Gjethedell
Reichardia picroides (L.) Roth		Herba dolça, cosconella	Tèrragrèpia, costelina	Lattarepulu	Caccialebbra	Lattaredda	Caccialepre, pane e cacie, sassello, Latticino	Galatsida	–	–	Buk, bukë ljepër	?
Rumex acetosa L.	Acedera	Agrella	Eigreto	–	–	Coraxedu	Zazzaio, pan e vino	Oxalidi	Avelyk, Trtndshyk	Stavelj	–	Lëpjetë
Scandix australis L.	Quijones	Agulloles	Agulhas	–	–	Erba de agullas	Acicula	–	–	–	–	Finraspor
Scorzonera laciniata L.	Burumbaya, verbaja, farfalla	–	Barbaboc, barbabou di champ, escorzonera, galineta	–	–	Crabiolina, pei de caponi	Scorzonera	Starida	Aishasindz	Zmijak	–	?
Sonchus tenerrimus L.	Lizones	Lisones, llicsó de perdiu	Lachasson	–	Cardedda	Alminzone, arminzone	Crispigno, cicerbita	–	–	–	–	–
Tamus communis L.	Támara	–	Tamisié	Sparaci	Sparacogni	Tamarit	Tamaro, Tàfani, dàfano	Avronia, ovria	Kyis armat	Bljust	Sparënx salvaç	Pejzë

Table 4b. Local GFP taxa in the Mediterranean in the Afro-Asiatic, Altaic and Caucasian Families of languages. Names are represented not only if the plant is used as food. Hebrew: [174–176]. Maltese: [177, 178]. Turkish: [79, 81, 179–183]. Arabic: [6, 84, 85, 115, 116, 184–196]. Berber: [115, 116, 185, 187]. Basque: [197–199, 236]. Azeri: [5]. Georgian: [5] (cognates and adoptions marked in shades of gray. See also equivalent rows in table 4a)

Taxon	Afro-Asiatic				Altaic		Caucasian	Isolate
	Hebrew	Maltese	Arabic	Berber	Turkish	Azeri	Georgian	Basque
Amaranthus graecizans L.	Yirebavooz yevanay	Cobbejra hamra	Fiss el kelb, lisan ter	Talenkhatait	Ohrasan	Gara tere	Dzhidzhilaka	–
Anchusa italica Retz. (= *A. azurea* auct.	Hallamut	Lsien il-fart ikhal	Bou-cassal, harcha	Foudelqqem, tament	Sigirdili, güriz	Siumiurgen	–	Belchoria, idimia
Arctium minus Bernh.	–	–	Lif el eulma	–	Uluavratotu	Pitrag	Orovandí	Lapa-txikia, Ooin, Orkatx
Beta maritima L.	Selek maavooy	Selq salvagg	Ders el kalb	Zmamûr	Yabani pankar	–	–	–
Eruca sativa L.	Orot	Aruka	Djergir, harra	Aoluet, tanekfayt	Roka	–	–	Berarakia
Eryngium campestre L.	–	–	Bou-adjoul	Oulouazene	Bogadikeni	Symbirtike	Lyrdji nari	Buruiska, chillarra
Lactuca serriola L.	Khaset hameetspez, merôrîm	Hassa salvagga tal-pizzi	Necalen, sacarole, sep-ghorab	Assafar n-sem	Marul, tahliç, yabani marul	Siuddien	–	Asta uraza
Lactuca viminea (L.) Presl.	–	–	–	–	–	–	–	–
Malus sylvestris Miller	–	–	–	–	Yabani elma	–	–	Sagar gacia

Table 4b. (continued)

Taxon	Afro-Asiatic				Altaic		Caucasian	Isolate
	Hebrew	Maltese	Arabic	Berber	Turkish	Azeri	Georgian	Basque
Muscari comosum (L.) Miller = *Leopoldia comosa* (L.) Parl.	Mtseeloovot mtsooyatsot	Basal il-hniezer	Beçol ed dib	–	Sümbül, morbas	–	–	Frailes
Myrtus communis L.	Hadas	Rihan	Rihân	Tarihant, tchilmoun	Banar, mersin	–	–	Arraiana
Plantago coronopus L.	Lechekh shasvoa	Salib l'asit	Bugenáhh, rjel el-gorâb	–	Kargaayagi	–	–	Izar belarra
Reichardia picroides (L.) Roth (& *R. tingitana* (L.) Roth)	Tamrooeyr maroqany	Kanklita	Zid et moum	Reghim	–	–	–	–
Rumex acetosa L.	–	–	–	–	–	Avelyk, gyzygylati	Golo	Gazigoxi, Miñeta
Scandix australis L.	–	–	Mechitta	Beghour, lebkour	–	–	–	Milluria
Scorzonera laciniata L.	–	–	Talma regiga	–	Karakök	Teke sajgali	–	–
Sonchus tenerrimus L.	–	Tfief	Zeizet el Maza	–	–	–	–	Kardabera, karduberacha
Tamus communis L.	Tamoos matsooey	–	Bou mimoune	Tsemimoune	Aciot, dövülmüsavratotu, karaasma	–	Dzaglis-satashyri	Asta matsa

languages from the common Proto-Indo-European language must have taken a long time, associated with the episodes of the original migration from Africa, followed by massive extinctions of Indo-European languages and the expansions of a few highly successful subgroups. The reconstructed Proto-Indo-European lexicon has a strong bias toward domesticated animals rather than crops [44, 49].

Diamond and Bellwood [49] see the Fertile Crescent not only as an agricultural homeland, but also as a zone that might have given rise to Indo-European, Elamite, Afro-Asiatic and Caucasian language families radiating from it. But it is unlikely that these expanded simultaneously to Neolithic agricultural diffusion. The homeland for Afro-Asiatic languages is also controversially placed either in the Levant during the late Natufian culture 9500 BC or in northeastern Africa.

Although there is debate on the origins of North and South Caucasian language families and as to whether they do or do not form a genetic unit, overall, the analysis of classical genetic markers is in agreement with a single ancient origin of Caucasian populations, followed by subdivision along geographical and linguistic lines [50].

It seems likely that the basic human genetic and cultural substrate of most of the Mediterranean Europe and North Africa was established in the region prior to the origins of yearly crop agriculture (cereals and pulses) and long before the end of the LGM. This allowed long term co-evolution – involving tolerance, resistance, domestication – between a wide repertoire of local food plant taxa and the existing human populations in the various glacial refugia.

Accordingly the different composition of the GFP lists and their local names (tables 2–4) could represent cultural differences between Eastern and Western Mediterranean human populations. Are these differences adaptive? Are these, not only cultural but also genetic, related with human genetic polymorphism, in function of types of food consumed? Differences in milk consumption and prevalence of lactase persistence in adults are the best known example but not the only one.

Plant Diversity

The largest of the world's five Mediterranean-climate regions, the Mediterranean basin stretches west to east from Portugal to Jordan and north to south from Northern Italy to Morocco. Around this sea is one of the most diverse world's biodiversity hotspots. Surrounding the Mediterranean Sea, the hotspot's 2,085,292 km^2 include parts of Spain, France, Italy, the Balkan states, Greece, Cyprus, Turkey, Syria, Lebanon, Israel, Egypt, Libya, Tunisia, Morocco and Algeria, as well as around 5,000 islands scattered around the Mediterranean Sea.

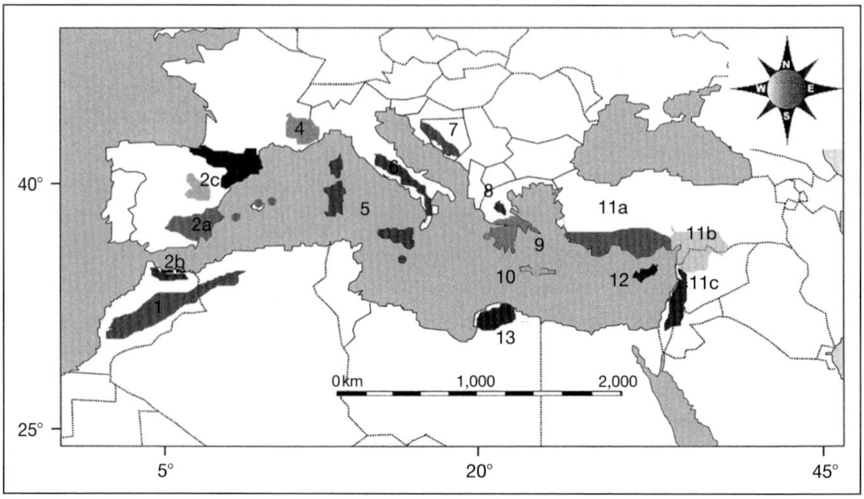

Fig. 3. Synthetic map of the 14 first micro hotspots of plant biodiversity in the Mediterranean basin: (1) the Atlas Mountains, (2a) the Baetic and Sub-Baetic range in Southern Spain, (2b) the Rif, (2c) massifs of Gudar, Javalambre and Serranía de Cuenca, (3) Pyrenees and neighboring ranges, (4) maritime and Ligurian Alps of the French-Italian border, (5) Tyrrhenian Islands, (6) Southern Appenins of Italy, (7) Dalmatian coast, (8) Mount Olympus, (9) mountains of Southern and Central Greece, (10) Crete, (11a) southwest Anatolia, (11b) Isaurian, Lycaonian and Cilician Taurus, (11c) Levantine uplands (Turkey, Syria, Israel, Jordan and Lebanon), (12) Troodos mountains in Cyprus, (13) Cyrenaica in Libya, and (14) the Canary/Madeira Islands (out of the range of the map) [Sources: 51, 52].

Of the 22,500 species of vascular plants in this hotspot, approximately 11,700 (52%) are found nowhere else in the world. The endemics are mainly concentrated on islands, peninsulas, rocky cliffs, and mountain peaks. Endemism at the higher level is highly limited. The Mediterranean only harbors two endemic families (Aphyllanthaceae and Drosophyllaceae), both represented by single species, *Aphyllanthes monspeliensis* L. (consumed as food in Spain) and *Drosophyllum lusitanicum* Link. On the other hand, the Mediterranean region is well known for its tree richness and a high degree of arboreal endemism (290 indigenous tree species with 201 endemics) [51].

The boundaries of Mediterranean are somewhat diffuse. The Alps to the North and the Caucasus harbor many Mediterranean species. The Sahara and the Irano-Turanian steppes merge with the driest Mediterranean areas.

The principal foci in the Mediterranean biodiversity are 14 regional mini-hotspots (fig. 3), characterized by areas of high plant richness and narrow endemism of >10%: (1) the Atlas Mountains in North Africa; (2a) the Baetic and Sub-Baetic range in Southern Spain; (2b) the Rif (two coastal strips of

Morocco and Algeria); (2c) massifs of Gudar, Javalambre and Serranía de Cuenca; (3) Pyrenees and neighboring ranges; (4) maritime and Ligurian Alps of the French-Italian border; (5) Tyrrhenian Islands; (6) Southern Apennines of Italy; (7) Dalmatian coast; (8) Mount Olympus; (9) mountains of Southern and Central Greece; (10) Crete; (11a) southwest Anatolia; (11b) Isaurian, Lycaonian and Cilician Taurus; (11c) Levantine Uplands (Turkey, Syria, Israel, Jordan and Lebanon); (12) Troodos Mountains in Cyprus; (13) Cyrenaica in Libya, and (14) the Canary/Madeira Islands. These 14 areas cover about 22% of the basin's total area, yet account for almost 5,500 endemic plants, i.e., about 47% of total Mediterranean endemics [51, 52].

Plant biodiversity includes also cultivated plants. With respect to crop plants, the concept of ethnovariety is very useful to tie regional agrobiodiversity to the general concept of local food. The idea takes cultural concepts about infraspecific diversity into account and an ethnovariety can best be defined as a useful taxon (mostly used for food) as it is understood and managed by local farmers, pastoralists or other users. Of course this includes most of the diversity of typical Mediterranean crops and fruits such as olive, grapevine, fig, broad beans, wheat, chards, cabbage as well as wheat, barley, chickpea, peas, oats and many others.

Importantly, the concept of agrobiodiversity includes not only the presently widely grown species but also, and especially, ones which have been almost abandoned and only remain as relict in some rural areas.

Cereal domestication first emerges in the Levantine corridor during the Pre-pottery Neolithic A period (ca. 10,700–9,300 BP) of the early 10th millennium BC and is then restricted to a few sites in the Southern and Central Levant for the next 400–500 years. Domestic cereals appear about 8700 BC, in southeastern Anatolia and on Cyprus in Early pre-pottery Neolithic B (ca. 9,000–8,500 BP) contexts. Cyprus is notable for representing the first definite evidence of a targeted migration by farming communities. Two or three centuries later, around 8500 BC, Central Anatolia was first settled by agricultural colonists. Finally, at the end of the 8th millennium, agricultural colonists arrived at approximately the same time in Southern Greece and Crete. A marked feature is the decreasing diversity, including weedy taxa, as crop packages moved from their areas of origin [53]. This is relevant because nearly 50% GFPs are weedy and were distributed together with the crop packages. GFPs richness at local level depends mainly on the plant diversity of an area. This led us to distinguish cryptocrops from weeds. Both are not cultivated plants living in crop fields and competing with the main crop. The fundamental distinction is the intensity of gathering by man. *Cryptocrops are gathered and used by local farmers in such way that they become a resource for them and a part of the local environmental management system. They are part of the TKS not only as a crude material but also as a tool for the complex management of secondary habitats*. The distinction is not taxonomic but ethnobotanical. One single taxon is a weed or a cryptocrop depending on the way it is managed. Many weeds were first used as a cryptocrop and then became a weed.

It is interesting to note that some weedy GFPs did not move beyond the Near East, such as *Gundelia tournefortii* L. (Compositae, Turkish *Kenger*, Arabic *Akkub*, Armenian *Kantchar*), *Bongardia rauwolfii* C. A. Mey. (= *B. chrysogonum* Boiss.) (Berberidaceae), or *Onopordum* spp. Others such as *Oenanthe pimpinelloides* L. (Umbelliferae, Greek *Sgarantsi*, Armenian *Dshenshajik*) are a more popular food in the Eastern Mediterranean. The presence in the area of GFP taxa since the Paleolithic period has been documented through archaeobotanical studies (fruits, pollen), for genera such as *Arctium*, *Centaurea, Crepis, Picris* or *Sonchus* [54].

The limits between wild and cultivated Mediterranean food plant species are often obscure. Hiort [55], Burg [56], Jerlin [57], Salberg [58], under the direction of Linnaeus, collected examples of, today often forgotten, cultivated taxa used as food plants from such diverse genera like *Campanula, Atriplex, Chenopodium, Rumex, Urtica, Borago, Sanguisorba, Sorbus, Bunium, Lathyrus, Smilax, Tamus, Arctium, Portulaca, Carlina, Onopordon, Reichardia, Scorzonera* and *Tragopogon.*

The Mediterranean Diets and GFP Consumption

The term 'Mediterranean diet' was coined in the book written by Ancel and Margaret Keys [59]. Later on, Keys demonstrated that traits, heretofore considered irrevocable and constitutional, such as body type, blood fat levels (cholesterol), blood pressure, heart rate, and responses to stress, were, in fact, largely modifiable by simple changes in the composition and quantity of diet and physical activity and that traditional Mediterranean diet helps to improve health parameters [60–62].

Naska [63] (modifying 64) defined nine components of Mediterranean diet: high consumption of olive oil and low consumption of lipids of animal origin – expressed in terms of a high ratio of monounsaturated to saturated fat; high consumption of vegetables; high consumption of legumes; high consumption of cereals (including bread); high consumption of fruit; moderate to high consumption of fish; low to moderate consumption of milk and dairy products; low consumption of meat and meat products; moderate wine (alcohol) consumption. One group ignored both by Naska [63] and Trichopoulou et al. [64] are local food plants, especially GFPs. The different local traditional Mediterranean diets include local food plants (especially GFPs) and in many cases little or nothing is known about their role in these diets and their health-beneficial effects [65–67].

Noah and Truswell [65] distinguished various groups of diets in Mediterranean countries (tables 1a, 7):

(i) *Western*: Spain, France, Italy and Malta. Consume bread, rice, pasta and potatoes as staple food, with vegetables and legumes. Cheeses are the most important item among dairy products. Olive oil is consumed mainly in Italy and Spain. Pork is the most important meat.

(ii) *Adriatic*: Croatia, Bosnia and Albania. Has a high consumption of white wheat flour as bread, also pitta. Dairy product consumption is high (butter, buttermilk, ricotta, cheese, sour cream, etc.). Olive oil consumption is low to moderate. Beef is the most important meat.

(iii) *Eastern*: Greece, Lebanon, Cyprus, Turkey and Egypt. All consume white flour products. Different sorts of cheese play an important role. Okra dishes are substantial in summer. Dill, parsley and oregano are essential herbs. Olive oil consumption is very high in Greece and negligible in Egypt. Chicken is important in this group.

(iv) *North African*: Libya, Algeria, Morocco and Tunisia. Eat bread made from whole meal flour and barley flour. Couscous is eaten more than rice. Potato, pumpkin and chickpeas are major foods. Dates and date molasses are essential for this group's diet. Olive oil consumption ranges from high to low. Lamb is the most used meat.

Bordering the Mediterranean are the Alps, Caucasus and Sahara. Alpine diet shows close resemblance with the Adriatic group (table 7) with a high consumption of white flour and dairy products. Low consumption of olive oil is a rule in the Alps. Beef is the most important meat although pork is also common.

The Caucasus' diet, far from the typical Mediterranean, included yogurt and garlic on a daily basis – they chop garlic and add it to yogurt in a dish called 'Sarimsagli gatig', used as a condiment. Meat was eaten only in the winter while green vegetables were primarily eaten in summer. Also consumed are fruits, including local apples. 'Bahmaz' (a concentrate made from white mulberries) was widely used [68, 69]. Traditional Caucasian diet (while being slightly variable it was shared by most ethnic groups of the region) is rich in milk and dairy products (matsoni – fermented milk, sour milk, butter, butter oil, cream, sour cream) compared with a standard Mediterranean diet. Other special features are high level of consumption of saffron, narsharab – boiled pomegranate juice, and cherry plums (fresh, dried or in form of dried thin sheets called lavashana) or doshab – boiled grape juice with added clay. It includes also GFPs, especially greens (tables 8, 9). GFPs are popular items in traditional recipes of Azerbaijan cuisine [70].

GFPs are present in many traditional dishes, table 6b records a repertory of traditional dishes with GFPs – some with ritual and religious meaning – in countries around the Mediterranean. Examples of the general use of GFPs are the

Table 5. Comparison of GFP diversity in 64 samples from Mediterranean countries and their boundaries. Differences in the area and the intensity/strategy of research determine significantly the numbers below. Ethnobotanical and ethnopharmacological studies are the basic sources. CI (% of taxa present in >20 samples/% of taxa in <5 samples*100)

Zones	Countries	GFPs and mushrooms	% exclusive (1 sample)	% (<5 samples)	% widespread (>20 samples)	CI	Area (1,000 km²)	References
Lepushe Alps	Albania	32	3	6	31	516.7	0.6	[238]
Armenia	Armenia	332	0*	43	10	23.3	29	[5, 200, 258]
Alps	Austria, France, Germany, Italy, Slovenia, Switzerland	358	24	52	10	19.2	290	[87–92, 120, 146, 214–218]
Azerbaijan	Azerbaijan	227	0*	39	12	30.8	86	[5, 243]
Istro-Romanians of Zejane	Croatia	14	0	0	42	–	1	[239]
Cyprus	Cyprus	129	13	34	17	50.0	9.2	[77, 171, 212]
Egypt	Egypt	60	27	67	3	4.5	997	[202–205]
Corsica	France	75	11	29	29	100.0	8.6	[123–223]
Provence	France	126	6	23	24	104.4	50	[91, 110, 214]
Savoie	France	142	20	50	8	16.0	3	[106]
Georgia	Georgia	272	0*	39	11	28.2	69	[5]
Caucasus	Georgia, Armenia, Azerbaijan, Russia	565	16	59	7	11.9	200	[5, 200]
Crete	Greece	200	18	46	15	32.6	8.2	[78, 200, 220, 256, 257] Z. Kypriotakis [pers. commun.]

Greece	Greece	168	10	32	20	62.5	132	[245–248, 254]
Iraq & Kurdistan	Iraq, Kuwait, Iran	88	22	55	7	12.7	450	[189, 190, 196, 200]
Palestine & Israel	Israel, Palestine	114	25	55	13	23.6	190	[84, 253]
Arbëreshë Vulture	Italy	68	4	10	46	460.0	1	[169]
Castelmezzano	Italy	63	5	16	38	237.5	1	[230]
Central Italy	Italy	111	3	17	27	158.8	3	[153]
Friuli	Italy	59	22	36	17	47.2	1	[221]
Galliciano	Italy	38	0	8	45	562.5	0.2	[231]
Sardinia	Italy	257	14	35	17	48.6	24	[124]
Serchio River (Carfagnana)	Italy	130	3	18	24	133.3	0.5	[235]
Sicily	Italy	71	14	34	23	67.6	25	[120–122]
Upper Lucca	Italy	68	0	9	38	422.2	0.4	[237]
Jordan	Jordan	195	27	59	10	16.9	89	[242, 250, 251, 261, 262]
Sahara	Libya, Algeria, Morocco, Tunisia, Mali, Niger	158	49	87	1	1.1	9,000	[6, 115, 116, 184, 186, 200]
Morocco	Morocco	225	28	56	11	19.6	450	[6, 115, 192, 200]
Portugal	Portugal	114	1	12	26	216.7	92	[4, 241, 252]
Arabia, UAE & Oman	Saudi Arabia, Oman, UAE	110	65	85	4	4.7	2,300	[188, 191, 200]
Albaida & Quatretonda	Spain	96	2	16	29	181.2	1	[229]
Alcaraz Segura	Spain	164	4	34	18	52.9	3	[72, 222]

Table 5. (continued)

Zones	Countries	GFPs and mushrooms	% exclusive (1 sample)	% (<5 samples)	% widespread (>20 samples)	CI	Area (1,000 km^2)	References
Almería	Spain	48	0**	40	29	72.5	8.8	[158]
Balearic	Spain	38	11	21	26	123.8	5	[153, 165, 224]
Basque country	Spain	111	5	14	27	192.9	7	[232–234, 236]
Cabañeros	Spain	61	2	14	30	214.3	1.5	[222]
Extremadura (East)	Spain	52	8	25	27	108.0	1	[147]
Huerta Murcia	Spain	63	8	30	27	90.0	0.4	[144, 163]
Madrid	Spain	117	5	22	30	136.4	8	[143, 161, 228]
Mágina	Spain	87	8	23	26	113.0	0.4	[160]
Marina Alta	Spain	43	5	19	35	184.2	0.4	[152]
Marina Baixa	Spain	33	9	21	33	157.5	0.5	[259]
Montseny	Spain	78	4	8	35	437.5	0.2	[111]
Navarra	Spain	104	0	13	25	192.3	10	[198]
Picos de Europa	Spain	134	11	28	22	78.6	4	[156]
Rest of Albacete	Spain	97	0	22	25	113.6	11	[9, 72, 222]
Rest of Cuenca	Spain	62	0	15	24	160.0	12	[9, 72, 222]
Segovia	Spain	64	3	20	31	155.0	6.9	[148]
Serranía Cuenca	Spain	147	1	18	22	122.2	5	[9, 72, 222]
Spain	Spain	562	12	44	8	18.2	504	[4, 240]
Valencia	Spain	120	10	23	25	108.7	25	[166, 226, 227, 260]
Vega Baja	Spain	30	7	23	40	173.9	0.8	[157]
Pyrenees	Spain, France, Andorra	199	6	25	19	76.0	40	[91, 145, 149, 151, 162, 225]
Canary & Madeira	Spain, Portugal	166	48	62	10	16.1	8	[206–211]
Syria	Syria	173	6	43	16	37.2	184	[85, 249]

Tunisia	Tunisia	125	14	50	14	28.0	160	[6, 200]
Ankara	Turkey	50	16	46	24	52.2	10	[182]
Eastern Anatolia	Turkey	177	29	58	11	19	200	[181]
Ilica, Erzurum	Turkey	42	26	43	17	39.5	1	[180]
Kizilkaya Aksaray	Turkey	94	24	49	19	38.8	3	[79]
Mugla	Turkey	142	16	46	16	34.8	1	[81]
Turkey	Turkey	407	18	54	9	16.7	780	[200, 213, 219, 244]
Western & Central Anatolia	Turkey	136	10	40	18	45.0	15	[82, 183, 200]
Western Sahara	Western Sahara, Morocco	38	16	79	0	0	97	[201]

*All taxa from Armenia, Azerbaijan and Georgia are included in Caucasus.
**Id. Almeria in Spain.

Table 6. Examples of Mediterranean endemic GFPs – those extending over different contiguous areas are cited in their main area (references as in table 5)

Zone	Greens, flowers, bulbs and tubers	Fruits and grains	Spices, herbs and drinks	Total
Spain	*Andryala integrifolia* L., *Aphyllanthes monspeliensis* L., *Asparagus albus* L., *Asparagus acutifolius* L., *Brassica cossoniana* Boiss. & Reuter, *Bunium macuca* Boiss., *Carduncellus dianius* Webb, *Carduus bourgaeanus* Boiss., *Carthamus lanatus* L. ssp. *baeticus* (Boiss. & Reuter) N. Br., *Centaurea diluta* Aiton, *Conopodium bourgaei* Coss., *Crocus carpetanus* Boiss. & Reuter, *C. nevadensis* Amo, *Cynara baetica* (Sprengel) Pau, *Diplotaxis crassifolia* (Raf.) DC., *Fumaria agraria* Lag., *Lactuca tenerrima* Pourr., *Leontodon taraxacoides Pau ssp. longirostris* Finch & P.D.Sell, *Linaria hirta* Moench, *Mantisalca salmantica* Briq. & Cavill., *Merendera pyrenaica* (Pourr.) P. Fourn., *Onopordon uniflorum* Cav., *O. nervosum* Boiss., *Reichardia tingitana* (L.) Roth., *Salvia argentea* L., *Scorzonera*	*Berberis hispanica* Boiss. & Reut., *Corylus hispanica* Miller, *Crataegus orientalis* M.Bieb., *Malus segurensis* Rivera et al., *Pinus pinea* L., *Pyrus bourgaeana* Decne., *P. cordata* Desv., *Quercus ilex* ssp. *ballota* (Desf.) Samp.	*Satureja intricata* Lange, *S. obovata* Lag., *Thymus hyemalis* Lge., *T. piperella* L., *T. zygis* L., *Ziziphora aragonensis* Pau, *Z. hispanica* L.	50

	angustifolia L., *S. crispatula* Boiss. ex Willk., *Sedum sediforme* (Jacq.) Pau subsp. *dianium* O.Bolòs., *Silene latifolia* Poiret, *Sisymbrium crassifolium* Cav., *S. erysimoides* Desf., *Sonchus crassifolius* Pourret ex Willd., *Stipa tenacissima* L., *Trachelium caeruleum* L.			
South France and Italy	*Bellevalia romana* Sweet, *Crocus neapolitanus* Mord. et Lois., *Gentiana kochiana* Perr. et Songeon, *Ornithogalum pyramidale* L.	*Quercus cerris* L., *Quercus virgiliana* Tenore	*Origanum heracleoticum* L., *Thymus pulegioides* L.	15
Western Mediterranean Islands	*Allium commutatum* Guss., *Asphodeline lutea* (L.) Reichnb., *Borago pygmaea* (DC.) Chatter & Greuter, *Brassica fruticulosa* Cyr., *Capparis orientalis* Veill., *Carlina hispanica* Lam., *Chamaerops humilis* L., *Crepis bursifolia* L., *Crocus minimus* DC., *Cuscuta epithymum* ssp. *corsicana* (Yuncker) Lambinon, *Hyoseris radiata* L., *Hypochoeris neapolitana* DC., *Lactuca longidentata* Moris, *Leontodon tuberosus* L., *Linaria reflexa* (L.) Desf., *Morisia monanthos* (Viv.) Asch., *Scolymus grandiflorus* Desf., *Scrophularia trifoliolata* L., *Tolpis quadriaristata* Biv., *Vinca sardoa* (Stearn) Pignatti	*Crataegus azarolus* L., *Pyrus amygdaliformis* Vill., *Ribes sandalioticum* (Arrigoni) Arrigoni	*Crocus longiflorus* Raf., *Mentha requienii* Bentham, *Ruta corsica* DC., *Teucrium marum* L., *Thymus herbabarona* Lois.	30

Table 6. (continued)

Zone	Greens, flowers, bulbs and tubers	Fruits and grains	Spices, herbs and drinks	Total
Balkans and Eastern Med. Islands	*Allium bourgeaui* Rech. f., *Allium neapolitanum* Cyr., *Anthemis melanolepis* Boiss., *Brassica cretica* Lam., *Bunium ferulaceum* Sibth. & Smith, *Campanula pelviformis* Lam., *Centaurea idaea* Boiss. & Heldr., *C. raphanina* Sibth. & Sm., *C. spruneri* Boiss. & Heldr. ssp. *minoa* (Heldr.ex Boiss.) Dostál, *Cichorium spinosum* L., *Conringia persica* Boiss., *Cynara cornigera* Lindley, *Echinops spinosissimus* Turra, *Hypochaeris cretensis* Benth. & Hook.f., *Muscari spreizenhoferi* (Heldr.ex Osterm.) Wehrh., *Onopordum cyprium* Eig, *Ornithogalum creticum* C.A.Zahariadi, *O. nutans* L., *Petromarula pinnata* (L.) A. DC., *Pimpinella peregrina* L., *Rumex thyrsiflorus* Fingerh., *Scaligeria napiformis* Grande, *Scariola acanthifolia* (Willd.) Soják, *Scorzonera cretica* Willd., *Smyrnium creticum* Mill., *Steptorhamphus tuberosus* (L.) Grossh., *Symphytum bulbosum* Schur., *Taraxacum aleppicum* Dahlst., *T. bithynicum* DC., *T. hellenicum* Dahlst., *T. megalorrhizon* Hand.-Mazz.	*Arbutus andrachne* L., *Prunus cocomilia* Ten., *Pyrus syriaca* Boiss., *Quercus macrolepis* Kotschy	*Origanum dubium* Boiss., *O. onites* L., *Salvia fruticosa* Mill., *Sideritis perfoliata* L.	60

Turkey and the Levant	*Allium atroviolaceum* Boiss., *Atriplex nitens* Schkuhr., *Arum conophalloides* Schott., *A. elongatum* Steven., *Asparagus palaestinus* Baker, *Beta lomatogona* Fisch. & Mey., *B. macrorhiza* Stev., *B. trigyna* Waldst. & Kit., *Bongardia chrysogonum* (L.) Spach, *Caralluma aaronis* (Hart.) NE. Brown, *Centaurea depressa* Bieb., *Crocus ancyrensis* (Herbert) Maw., *C. hyemalis* Boiss., *Cynara syriaca* Boiss., *Echinophora orientalis* Hedge & Lamond, *Echinops pungens* Trautv., *Eremurus spectabilis* Bieb., *Gladiolus atroviolaceus* Boiss., *Iris galatica* Siehe, *Jurinea pontica* Hausskn., *Leontice leontopetalum* L., *Merendera kurdica* Bornm., *Orchis simia* Lam., *Rheum ribes* L., *Salvia candidissima* Vahl., *S. hyerosolimintana* Boiss., *Scorzonera cana* (C.A. Meyer) Hoffm., *Sempervivum armenum* Boiss. et Huet, *Tragopogon buphtalmoides* (DC.) Boiss.	*Amygdalus orientalis* Miller, *Berberis crataegina* DC, *Corylus* spp., *Crataegus aronia* (L.) DC., *C. meyeri* Poyark., *C. pontica* C. Koch, *Pistacia atlantica* Desf., *Prosopis farcta* (Banks & Sol.) JF Machr., *Prunus divaricata* Ledeb., *P. ursina* Kotschy, *Quercus infectoria* Olivier, *Q. ithaburiensis* Decne., *Q. pubescens* Willd., *Q. trojana* P.B. Webb. *Ziziphus spina-christi* (L.) Desf.	*Salvia hypargeia* Fisch. Et Mey., *Satureja thymbra* L., *Thymus fallax* Fisch. & Mey., *T. longicaulis* C. Presl., *T. sipyleus* Boiss., *Ziziphora clinopodioides* Lam.	80

Table 6. (continued)

Zone	Greens, flowers, bulbs and tubers	Fruits and grains	Spices, herbs and drinks	Total
Caucasus	*Biebersteinia multifida* DC., *Lilium* spp., *Michauxia laevigata* Vent., *Papaver orientale* L., *Spinacia tetrandra* Stev.	*Corylus* spp., *Diospyros lotus* L., *Fagus orientalis* Lipsky, *Malus orientalis* Uglizk., *Pyrus* spp., *Prunus* spp. *Vitis vinifera* L. *ssp. caucasica* Vav.	*Dryas caucasica* Juz., *Quercus iberica* Bieb., *Satureja macrantha* C. A. Mey, *S. mutica* Fisch. & Mey., *Staphylea pinnata* L., *S. colchica* Stev., *Ziziphora tenuior* L.	80
North Africa and Sahara	*Aizoon canariense* L., *Anchusa aegyptiaca* DC., *Arum hygrophyllum* Boiss., *Asphodelus tenuifolius* Cav., *Atriplex halimus* L., *Brassica tournefortii* Gouan, *Caralluma europaea* ssp. *gussoneana* (Mik.) Maire, *Cymbopogon schoenanthus* (L.) Spreng., *Cynara baetica* (Sprengel) Pau ssp. *maroccana* Vicklund, *Diplotaxis douveyreana* Coss., *D. harra* (Forssk.) Boiss., *Erodium guttatum* (Desf.) Willd., *E. glaucophyllum* (L.) L'Hér., *E. hirtum* (Forssk.) Willd., *Euphorbia granulata* Forssk., *Hedysarum carnosum* Desf., *Launaea nudicaulis* (L.) Hook., *Matthiola livida* (L.) D.C., *Myrtus nivellei* Batt. & Trab., *Rumex tingitanus* L., *R. thyrsoides* Desf., *R. vesicarius* L., *Salvadora persica* L., *Schouwia purpurea* (Forsk.) Schweinf., *Silybum eburneum* Cosson et Durieu	*Acacia albida* Del., *Aristida adscensionis* L., *A. pungens* Desf., *Asphodelus refractus* Boiss., *Astragalus lanigerus* Desf., *A. vogelii* (Webb) Hutch., *Dactylotenium aegyptiacum* Willd., *Lolium multiflorum* ssp. *gaudini* Schinz & Keller, *Nitraria retusa* (Forssk.) Asch. *Panicum turgidum* Forsk., *Pennisetum dichotomum* Forssk., *Pistacia atlantica* Desf., *Plantago ciliata* Desf., *Pyrus mamorensis* Trab., *Ranunculus macrophyllus* Desf., *Rhus tripartitum* (Ucria) DC., *Tamarix aphylla* (L.) Karst., *Ziziphus lotus* (L.) Lam., *Z. saharae* (Batt.) Maire	*Ammodaucus leucotrichus* (Coss. & Dur.) Coss., *Cymbopogon schoenanthus* (L.) Spreng., *Phlomis floccosa* D. Don, *Salvia aegyptiaca* L.	60

Table 7. Food patterns of Mediterranean countries. Data for Italy, Morocco, Lebanon, Turkey in percentage servings/head/week, Nicotera in % of g/head/day, Oldways' in % of total area of the Harvard's model of a 'Mediterranean' diet pyramid [adapted from 65]. Data for Alps [93] and Azerbaijan (Caucasus) [70] where frequencies are percentage of citation in the recipes for each ingredient. In bold those parameters that are significantly high

Food types	Western	Western	North African	Eastern	Eastern	Adriatic	Alpine	Caucasian	Oldways' diet
Area	Italy	Italy (Nicotera)	Morocco	Lebanon	Turkey	Croatia & Slovenia	Italy (Trentino and Dolomites)	Caucasus (Azerbaijan)	Nowhere
Bread, burghul & pasta	11.16	**35.82**	**14.38**	10.17	10.54	9.84	**16.60**	**13.50**	**26.83**
Vegetables	13.30	**25.74**	9.82	14.05	14.56	**17.62**	9.98	**19.02**	**20.33**
Fruits	**19**	7.47	**21.23**	15.71	17.24	11.89	7.98	13.50	13.01
Legumes	10.93	3.69	10.96	15.53	**13.60**	2.46	0.74	0.61	**13.82**
Milk, cheese & dairy products	5.70	4.14	7.31	8.87	8.62	**16.39**	**24.26**	**13.50**	7.32
Alcoholic beverages	3.09	**10.71**	0	1.48	1.53	1.23	5.57	4.91	0
Meat	**11.88**	3.06	6.16	4.99	4.60	**16.39**	**9.56**	**11.66**	0.81
Nuts	4.04	0	4.79	8.69	**9**	1.23	2.63	0	2.44
Sweets	2.61	1.98	3.20	3.51	3.64	**8.61**	5.04	12.88	0.81
Olive oil	**3.09**	**3.42**	0.68	1.48	1.53	1.23	1.68	0	**8.13**
Fish & fish products	1.90	**2.88**	**2.97**	1.48	0.57	1.23	1.58	0	**4.07**
Eggs	1.90	1.08	1.83	2.40	2.49	0	**8.61**	**9.20**	2.44
Coffee	3.09	0	2.97	2.40	2.49	**5.33**	0	0	0
Suet	1.90	0	**2.97**	2.40	2.49	1.23	2.10	0	0
Tea	1.90	0	**2.97**	1.48	1.53	**5.33**	0	0	0
Honey	1.90	0	**2.97**	2.40	2.49	0	0.32	0	0
Potatoes	1.90	0	**2.97**	1.48	1.53	0	**2.63**	1.23	0
Other vegetable oil	0.71	0	**1.83**	1.48	**1.53**	0	0.74	0	0

Table 8. Selected examples of traditional local dishes with GFPs in the Mediterranean and surrounding areas (sources as in table 5)

Dishes	Country	Comments
Turlu? (guevej)	Armenia	Mixed vegetable dish with baked lamb (a baked stew) (cooked)
Gananchehen	Armenia	Green vegetables (cooked)
Kutaby from wild-growing plants	Azerbaijan	Wild greens (cheese-flower, shepherd's purse, ziziphora, white dead nettle, chickweed, etc.) used as filling of a local type of 'calzone'
Salada Campanèla	France (Province)	Non-cultivated edible greens. *Cichorium intybus, Lactuca perennis, Campanula rapunculus, Plantago coronopus, Valerianella* sp., *Crepis bursifolia, Urospermum dalechampii* (cooked or raw)
Romiciata	France (Corsica)	Pastry filled with *Rumex* spp.
Suppa d'erbiglie	France (Corsica)	Soup made with a complex mixture of GFPs: *Allium ampeloprasum, Foeniculum vulgare, Reichardia picroides, Silene vulgaris, Sonchus* spp., and others altogether with carrots, onions, potatoes, beans. Is often prepared with rice or with bread crumbs
Xorta, xortopita	Greece	Autumn horta with edible *Chrysanthemum*, vinegar and olive oil
Pea shoots, papoules	Greece	Pea shoots (*Lathyrus ochrus*), tendrils and buds with tomato and onion
Melagria	Israel	Edible roots (*Asphodelus microcarpus*) the Christian Anchorites used to dig up in the Judaean Desert
Merorim	Israel	Bitter herbs and unleavened bread are served with the roasted lamb in the Passover. What kind of salad is intended by the word merorim, which literally signifies, 'bitters' is not well known. The Jews think: Dandelion *Taraxacum* spp., chicory *Cichorium* spp., lettuce *Lactuca* spp., watercress *Rorippa* spp., sorrel *Rumex* spp., arugula *Eruca* spp., akravnin *Scorpiurus* spp. The bitter herbs represent the bitterness of life under slavery
Broa	Italy (Trentino, Alps)	Mixed bitter vegetable dish with *Papaver* seeds and *Brassica rapa*
Pistic	Italy (Western Friuli)	A collection of 56 GFPs, which are boiled and then sauted together on occasion of a spring festival

Dish	Country	Description
Minestra delle 18 erbe selvatiche	Italy (Barbagia, Sardinia)	Mixed vegetable dish with *Borago officinalis, Silene vulgaris, Beta maritima, Carduus pycnocephalus, Sonchus arvensis, Diplotaxis muralis, Lathyrus pratensis, Papaver rhoeas, Rumex conglomerates, R. scutatus* and up to 18 different GFP taxa
Arugula	Italy	Arugula-smothered bean salad with prosciuto
Turione crudo	Italy (Sardinia)	Raw *Crepis vesicaria* sprouts with *Reichardia picroides* and *Rumex scutatus* salad
Idam	Libya (Fezzan)	Fat sauce with *Rumex vesicarius* and *Launaea glomerata*
Bqûla (edible wild greens)	Morocco	*Malva* spp., *Silene vulgaris, Portulaca oleracea, Papaver* spp. Non-cultivated greens parboiled and fried, served with preserved lemons and olives
Îggdîwen (edible wild greens)	Morocco	*Urtica* spp., *Silene vulgaris, Rumex* spp., *Papaver* spp. Non-cultivated greens parboiled and fried
Msâhen	Sahara	Complex seasonign mixture of wild hot herbs and spices gathered in the Sahara (*Androcymbium gramineum, Acacia raddiana, Centaurea pungens, Marrubium desertii, Eremophyton chevallieri, Limoniastrum guyonianum*)
Râs el-hanût	Sahara	Mixture of perfumed spices and wild greens of the Sahara used for seasoning a wide range of dishes
Ensalada del campo	Spain	Wild greens (*Sonchus* spp., *Papaver* sp., *Cichorium* spp., *Crepis vesicaria, Urospermum picroides*, etc.) salad dressed with vinegar and extra virgin olive oil
Cocas and Mintxos	Spain	Local type in Valencia and Alicante of the Italian calzone and pizza, filled with fish and wild greens (*Sonchus* spp., *Reichardia* spp., *Beta maritima* L., etc.)
Chekchoua	Tunisia	Dish with red peppers, beans and wild greens
Mhammes	Tunisia	Boiled greens: *Urtica pilulifera* L.
Mallow leaves	Tunisia	Mallow leaves with Tunisian spices
Mustard greens	Turkey	Mustard greens with black-eyed peas and rice

Table 9. Comparison of ingredient frequencies in traditional recipes from: Mediterranean [71], Trentino and Dolomites (Italian Alps) [93], Azerbaijan (Caucasus) [70], Castilla-La Mancha (Spain) (only those with GFPs) [72]. Values in percentage of recipes in which each ingredient is used. Years of co-evolution estimated from [73, 74]. (AMH = anatomically modern humans). In bold those parameters that are significantly high

Name	Years of co-evolution with Mediterranean AMH	Relative frequency Mediterranean cookbook	Relative frequency in Trentino and Dolomites	Relative frequency in Azerbaijan	Relative frequency in Castilla-La Mancha
Vinegar (grapes or pomegranates)	4,000	7.38	6.29	**20**	12.5
Wine	5,000	8.81	**17.71**	0	0
Bread	2,000	7.86	**74.29**	0	9.17
Rice	2,000	5.48	2.29	**25**	5
Wheat flour	8,000	7.14	**54.86**	**50**	5.83
Cheese	8,000	**14.76**	**25.71**	0	0
Eggs	50,000	7.62	**46.86**	**37.5**	**31.67**
Fish	50,000	**12.62**	**8.57**	0	2.5
Silene vulgaris	6,000	0	0	0	**14.17**
Wild fruits	50,000	0	**7.43**	**7.5**	**6.67**
Wild greens	50,000	0	1.14	**20**	**55.83**
Wild mushrooms	50,000	0.71	**5.14**	**0**	**32.5**
Beans	500	4.76	4	2.5	**11.67**
Chick peas	8,000	3.1	0	0	**10**
Chicken	2,500	**8.57**	0	5	2.5
Mutton and lamb	8,000	2.62	0.57	**42.5**	1.67
Pork	8,000	1.9	**35.43**	0	**18.33**
Butter	8,000	3.57	**73.71**	**55**	0
Extra virgin olive oil	6,000	**51.9**	9.14	0	**89.17**
Coriander	4,000	1.19	0	**32.5**	0.83
Flat leaf parsley	3,000	**22.14**	**20.57**	10	3.33
Garlic cloves	4,000	**33.1**	19.43	10	**38.33**
Ground black pepper	2,000	**38.81**	**40**	**55**	0
Hot red chili	500	7.86	0	0	0
Lemons	3,000	**22.86**	**15.43**	**20**	0.83
Sea salt	**50,000**	**94.05**	**76**	**67.5**	**88.33**
Sweet paprika	500	3.33	2.86	0	**13.33**
Sugar	1,200	8.81	**27.43**	**52.5**	2.5
Onions	3,000	**30.71**	**34.29**	**57.5**	15.83
Potatoes	500	7.62	14.29	5	**26.67**
Tomatoes	500	**22.62**	4	15	**22.5**
Water	50,000	20	20	**62.5**	**41.67**

Ganancheghen (Armenia), Kutaby (Azerbaijan), Salada Campanèla (Provence, France), Romiciata (Corsica, France), Xorta, xortopita (Greece), Merorim (Israel), Broa (Trentino Alps, Italy), Minestra delle 18 erbe selvatiche (Sardinia, Italy), Idam (Fezzan, Libya), Bqûla (Morocco), Msâhen (Sahara), Ensalada del campo (Spain).

There is relatively abundant information on GFP taxa consumed in the Mediterranean countries (table 2a). However, data on recipes and frequency of use are scarce. Table 9 presents a comparison of ingredient frequencies in a standard reference Mediterranean cook book [71], in a Caucasian (Azerbaijan) collection of 30 traditional recipes [70] and in the repertory of 200 traditional recipes with GFP species [72] in Castilla-La Mancha (Spain) with the purpose of analyzing the GFP consumption and its correlation with other foods and ingredients.

A high proportion of eggs and a low proportion of fish, milk and dairy products are found in the GFP recipes of Castilla-La Mancha. Fish is not easily available in the inland mountain areas of Castilla-La Mancha and this could explain its low prevalence, similar to the Azerbaijan's recipes. High egg frequency in the GFP recipes can be alternatively interpreted either as a cheap and accessible source of animal fat and proteins or as the use, more interesting, as traditional detoxifying agent by supplying the organism with cystine (two cysteines linked by a disulfide bond). Cysteine is a precursor of glutathione. The sulfhydryl group of cysteine is responsible for the chemical properties of the whole glutathione molecule (*L*-γ-glutamyl-*L*-cysteinylglycine). Hen ovalbumin contains one cystine disulfide and four cysteine sulfhydryl groups in a single polypeptide chain of 385 amino acid residues. Are omelettes and scrambled eggs a way of preventing harmful effects of toxic factors in foods making safer the consumption of GFPs?

Another interesting point is the use of water for cooking in much higher proportion (41% instead of 20%) in the GFP recipes.

Another relevant aspect is the high prevalence of ancient food types (estimated interaction time with Mediterranean anatomically modern humans ca. 35,000 years[1]) in the GFP recipes compared with the standard collection of recipes by Harmon in 1994 [71]. The standard recipes include in higher proportion post-Neolithic food types. However, and surprisingly, American food plants (potatoes, tomato, peppers) are much more frequent in the GFP recipes. Is this reflecting the adaptive value of the local traditional knowledge?

GFP groups seem still difficult to define but some major notes are found (table 2a). The 'meadow greens' that are basic ingredients of 'Salada campanela'

[1]Estimated interaction time with Mediterranean anatomically modern humans in years is based for each food type on different sources Olver [73] and Kiple and Ornelas [74].

and similar recipes in Romance-speaking countries of the Western Mediterranean characterize this western type. Although North Africa is not well defined by the lack of studies it seems that small wild grains (achenes, caryopses, etc.) altogether with underground parts (bulbs, rhizomes, corms) and ground-level buds and sprouts (e.g. *Phragmites*) are much more relevant than in the north. The eastern type is rich in tender cardoons (that indeed are also consumed in the west). The Caucasus is again rich in 'meadow greens', but the available information does not provide a clear picture of the Caucasian type.

Comparing Local Cultural, Genetic and Ethnobotanical Data

The Eastern Mediterranean

Theophrastus (4th century BC) described results of observation on wild-cultivated plant relationship in Greece. Concerning GFPs he stated: 'In some cases however the wild kinds are not in the least like the cultivated in taste and properties . . . ' He recognized their suitability for food whether raw or cooked [75].

No single plant family can be named as providing the staple food of classical Greece. Some of the poorest of country-dwellers may have relied regularly on acorns – as in many other areas around the Mediterranean Sea. Some pulses and cereals had a place in the diet that is now difficult to define for lack of evidence. Wild and cultivated vegetables were normally served in classical Greece as appetizers, side dishes and potherbs, even though those who always went for fish and meat despised them. Certain vegetables, the sources suggest, were food only for the poor. Examples include asphodel, wild chervil, chervil, a type of goosefoot, rocket, nightshade, sow thistle, etc. [76].

Savvides [77] published a selection of 66 taxa of GFPs based 'on the existing tradition of Cypriot countrymen, who have been used for thousands of years as a source of food' but also on the bibliography and on the author's personal life experience on this subject. This list of Cyprian GFPs contains mainly founder species of the complex of early domesticates.

Kypriotakis [pers. commun.] recorded ca. 150 different GFP taxa in Crete. Popular accounts of local food plants and recipes from Crete and other Mediterranean islands are growing in number steadily. Lambraki [78] recorded 100 more or less traditional recipes and 30 beverages in Crete (Greece), based in personal experience and bibliographic study.

Although Central Anatolia (Turkey) is at the periphery of the Mediterranean area, the ethnobotanical study carried out by Ertug [79] furnishes an interesting and detailed recording in some representative small Turkish villages among the non-Indo-European language-speaking population – with low Central-Asian genetic background [41]. The list includes 133 species of GFPs belonging to 31 plant families. Fungi are poorly represented (only 7 species). Further information is recorded for Samsun [80] and Bodrum areas [81]. Dogan et al. [82] studied 121 wild edible plants, used as food, boiled, fried in fat, and eaten raw or as rolled vegetables, in Eastern Anatolia. They are also consumed as pickles, fruits, sweets and spices, and drunk as cold and hot drinks. Emphazis is given to condiments (flavoring ingredients): 30 Labiatae species, 15 Compositae, 13 Rosaceae, 8 Cruciferae, 6 Orchidaceae and 5 Umbelliferae.

The Levant

In the Medieval Islamic Mediterranean cultures the borderline between food and medicine appears extremely diffuse. Huici [83] recorded many recipes of 'functional foods', viz. dishes especially recommended for pregnant women, persons with liver complaints, etc. Still in the 20th century complex mixtures of herbs and spices were used in Palestine and Syria that were consumed as a tea or beverage for improving health [84, 85].

Content to survive solely on wild flowers and plants, such Anchorites (an early monastic movement of the pre-Byzantine period) are known in Greek literary sources as *Boskos*. The term has a dual meaning: it is the word for shepherd, but it can also be employed for describing one whose sole source of nourishment are wild plants (grazers). The Syrian term *raaia* has a similar dual meaning when used in reference to anchorite monks. In the Judaean desert many Laurite monks, during periods of wanderings, they lived as *boskoi*, and their sole source of food were the plants growing wild in the desert (such as Melagria, *Asphodelus microcarpus*) [255].

In Jordan [86], 56 wild edible plants are used in local meals either eaten raw, cooked by boiling in water, fried in fat or baked and the GFPs are then served as dishes such as stews, turnovers, stuffed and rolled vegetable foods or as cold or hot drinks for certain occasions and seasons.

The Alps

The Alps extend southwest to east from France, through Switzerland, Italy, Germany, Austria, Slovenia to Hungary. They are an important refuge for cultural and biological diversity. Linguistic isolates are frequent especially in the valleys facing south (Ladin speakers in the Dolomites, French, Romanche and Walser speakers in Val d'Aosta, South Bavarian, German and Cimbrian speakers in Trentino and South Tyrol). GFPs have been recorded in different areas: Tyroler Alps [87, 88], Aosta [89], Dolomiti [90], Dauphiné's Alps [91, 92]. The high wild mushroom consumption and diversity is a typical Alpine feature. Nearly 250 different taxa of fungi are sold in the market held from June until November in Trento (Italy) [93]. Nuts, fresh fruits and meadow greens define the Alpine GFP type. It is also found in the Pyrenees, Cantabrian range and the Caucasus (table 2a).

The Caucasus

Recent ethnobotanical studies in the Caucasus are lacking but Grossheim [5] recorded an exhaustive catalogue of the plant resources (wild and cultivated, food and non-food). GFPs account over 500 for the whole Caucasus (table 5) and its regular consumption has been invoked as a reason for centenarians living in Georgia and Azerbaijan. Claims – with no reliable documentation [94] – that some of the oldest people on the earth live in the foothill regions of Caucasus exist.

The Caucasus exhibits a high degree of linguistic diversity, with four linguistic families (North Caucasian, South Caucasian, Indo-European, and Altaic) spoken by ca. 50 autochthonous groups. Neither geographic nor linguistic relationships appear to explain the genetic relationships of Caucasus populations. Instead, it appears as if they have been small and relatively isolated, and hence genetic drift has been the dominant influence on the genetic structure of Caucasus populations [95]. Here Indo-European-speaking Armenians and Turkic-speaking Azerbaijani are genetically most closely related (for both mtDNA and the Y-chromosome) to other Caucasus groups and not to other Indo-European or Turkic-speaking groups. North and South Ossetians that speak an Indo-European language have a common origin from Iran [96, 97]. Discordance between languages and genes may arise when an expanding language is imposed on or adopted by a peripheral population, with only a minor contribution of expanding genes. A possible historical example is the imposition of the Turkish language on medieval Anatolians (and Caucasians), whose genes today are estimated to be derived only 30% from their conquerors' genes [49].

Overall, the Caucasus groups showed greater similarity with West Asian than with European groups, although this similarity was much more pronounced for the Y-chromosome than for mtDNA [98]. To explain why mtDNA places the Caucasus in an intermediate position between Europe and West Asia, Nasidze et al. [98] suggest that this reflects a common ancestry of Caucasus and European populations dated back to the pre-Neolithic times [99].

The Western Mediterranean Europe

The Roman naturalist, Pliny, in his Natural History (1st century AD), refers to many local Mediterranean food plants either as local ethnovarieties of cultivated plants or GFPs. Many are interesting examples of shared Mediterranean local foods. The Greek myrtle (*Myrtus communis* L.) widely grown and consumed in Rome. The grapevine local cultivars in different areas of the Mediterranean suggest the setting of a geographical mosaic for Mediterranean agrobiodiversity in Pliny's time. The acorns, consumed as food and in the Spanish provinces finding a place in the second course at table, point to the prevalence of food types now almost abandoned. The truffles, whose African variety is the most highly spoken of, introduces us to a highly selective gastronomy. The wild and cultivated bulbs are thoroughly discussed in the Pliny's work. Thistles are cited as a most profitable article of trade at Carthage and Cordova (*Scolymus* and *Cynara* GFPs). The excellent qualities of wild sorrel or the chicory or Egyptian wild endive are described [100–102]. A careful analysis of Pliny's Natural History would give us a novel perspective for present-day ethnobotanical data. However, the exact botanical identity of the plants he cites remains a serious problem. Österman [103], under the auspices of Linnaeus, provided an interesting analysis of the foods, particularly plants, consumed by the Roman and Greeks: with GFPs such as acorns, mallows, amaranths, borage, orach, asparagus, arugula, and others.

Ignacio Jordán de Asso [104] recorded in his synopsis some GFPs collected in Aragon (Spain) (table 10). He also refers for uses and names to a previous manuscript by Cienfuegos (17th century) [105]. Chabert [106] comments on the problems experienced by the French botanist Villars relating to his 'Catalogue of Food Plants' [107]. The reputed scientist compiled an exhaustive list of wild food plants (including famine food) of French Dauphiné, whose publication was badly received and led him to resign his chair as professor of botany in Grenoble. Under the French Revolution, famine threatened France but people at the major cities disliked the consumption of wild vegetables as an alternative for food and especially bread shortage. Notwithstanding, the diversity of local food plants was high in the times of Villars and still in those of Chaubert [106]. Gaut [108] described a Provençal country salad comprising a high number of species: *Rumex* spp., *Apium graveolens, Tragopogon pratense, Silene vulgaris, Portulaca oleracea, Urospermum picroides, Centaurea cyanus, Scorzonera laciniata, Sanguisorba minor* and many others. This is reported by Réguis [109] under the names of 'Salado campanello' or 'Saoutourno'. Chauvet et al. [110] focused again on over 50 species of greens traditionally consumed in Occitan areas (southeastern France) under the name of 'Ensalada campanèla' or field salad. This idea of a wide diversity of salad plants collected from the fields but including exclusively non-cultivated plants (wild and weeds) is found repeatedly in the Western Mediterranean. Thus in Murcia (Spain) we found again under 'Ensalada del campo' and 'Ensalada de la huerta' an extraordinary compendium of GFPs similar to those recorded in Languedoc and Provence and in Valencia as 'Camarotges'.

Bonet and Vallès [111] in an ethnobotanical survey carried out in the massif of Montseny (Catalonia, Spain), recorded the different ways of preparation, preservation and consumption of 132 GFP taxa.

Table 10. Local GFPs in Aragon (Spain) recorded by Asso [104]

Plants and fungi	Local name	Data on uses	Family
Chenopodium bonus-henricus L.	Serrones	Plant consumed	Chenopodiaceae
Eryngium campestre L.	Cardo corredor	Roots sweet, eaten	Umbelliferae
Bunium bulbocastanum L.	Bulbocastaña	Roots edible	Umbelliferae
Rumex scutatus L.	Acedera	Tender leaves consumed, and sold in the market of Zaragoza	Polygonaceae
Rumex acetosa L., *R. acetosella* L.	–	Tender leaves edible	Polygonaceae
Arbutus unedo L.	Madroño	Fruits edible	Ericaceae
Portulaca oleracea L.	Verdolaga	Consumed in salads	Portulacaceae
Lepidium sativum L.	Mastuerzo	Consumed in salads	Cruciferae
Scorzonera laciniata L.	Farfallas	Leaves edible	Compositae
Chondrilla juncea L.	Achicorias dulces	Consumed in salads in Spain	Compositae
Cichorium intybus L.	Achicorias amargas	After boiling leaves consumed in salads	Compositae
Scolymus hispanicus L.	Cardillos	It is eaten in Spain, boiled with meat (the leaf rachis?)	Compositae
Sanguisorba minor Scop.	Pimpinela	Consumed in salads called 'Ensalada Italiana'	Rosaceae
Bryonia cretica L. *ssp. dioica* (Jacq.) Tutin	Nueza blanca	Sprouts consumed in omelets	Cucurbitaceae
Celtis australis L.	Almez, lodoño	Children eat the fruits	Ulmaceae
Carlina acanthifolia All.	Cardo aljonjero	Receptacle consumed	Compositae
Calocybe gambosa (Fr.) Donk	Cagarrias	Edible, delicious	Agaricaceae
Terfezia claveryi Chatin	Criadillas de tierra, turmas	The largest in Spain come from Murcia but those smaller are also delicious	Terfeziaceae

In Western Europe during the LGM, the Iberian Peninsula was a refuge for plants and humans. Here demographic factors associated with contraction, isolation, subsequent expansion and gene flow episodes have contributed complexity to its population history of a limited heterogeneity in the region. These results indicate that neither old or recent Levantine expansions nor North African contacts have influenced the current Iberian Y-chromosome diversity. Since the colonization of the Iberian Peninsula by modern humans (13,000–25,000 BC) the evolution of the Cantabrian area, in particular Galician and Basque country populations, seems to be genetically very homogeneous [112].

Mitochondrial DNA analysis of Atlantic European samples detected significant latitudinal clines for several clusters. An important gene flow from Africa was detected in Atlantic part of the Iberian Peninsula. Specific sub-Saharan lineages appeared mainly restricted to Southern Portugal, and could be attributed to historic Black slave trade in the area and to a probable Saharan Neolithic influence. Haplotypes of specific North African origin have only been detected in the Iberian Peninsula northwards from Central Portugal. González et al. [113] reject the proposal that only historic events such as the Moslem occupation are the main cause of this gene flow, and instead proposed a pre-Neolithic origin for it. However, Comas et al. [114] (analyzing insertion polymorphisms in several northwestern African and Iberian populations) show a clear differentiation of northwestern African and Iberian groups of samples, suggesting a strong genetic barrier matching the geographical Mediterranean Sea barrier. Some degree of gene flow from sub-Saharan Africa can be detected in the southern part of North Africa and in Saharawi and southeastern Moroccans, as a result of a continuous gene flow across the Sahara desert.

North Africa and the Sahara

Bellakhdar [115] includes an exhaustive analysis of GFPs in Morocco within his monograph on medicinal plants. He included not only medicinal GFPs but also those consumed as food without any known medicinal use. Gast [116] exhaustively reported on 80 species of wild vegetables and grains (ca. 20% of the vascular flora) of famine food plants of the Central Sahara in the Ahaggar region occupied by Berber groups. Seasonal food shortage leads to a famine period from December to March. North African and Saharan types of GFPs coincide in the Maghreb countries but are markedly distinct. The Saharan type markedly excludes most of the commonest GFP taxa that occur in the Mediterranean. Those present in >33% of samples in table 5 are not consumed in the Sahara.

Nebel et al. [117] show closest relationships between northwestern Africa, the Southern Levant and Yemen studying the frequencies of Y-chromosome Eu10 haplotypes. Mitochondrial DNA transit between West Asia and North

Africa has been inferred from U6 haplotype phylogeography. It spread to North Africa from the Near East around 30,000 BP [118].

In general the refugia during the LGM acted here not only for humans but also for GFP species that in warmer periods spread northwards to areas that became free of ice and southwards in the areas left by the recession of the Sahara desert. This suggests a Western Mediterranean and North African refuge area for wild greens and fruits in a similar way as for humans. This included species such as *Scorzonera hispanica* L., *Papaver setigerum* DC, *Pinus pinea* L., *Quercus ilex* L., and others.

Mediterranean Isolates: Outliers and Islands

Giuseppe Pitre in his Sicilian Folk Medicine noted that in Sicily the people eat just enough to enable them to work and to live, their food being basically vegetables (wild and cultivated) with low consumption of meat [119]. Presently, consumption of GFPs is particularly common, in particular near the Etna, in Catania [120–122].

The soup of wild herbs [123] is a typical Corsican dish with variable recipes, comprising a wide repertory of weeds and meadow greens: especially *Reichardia picroides* Roth but also *Silene vulgaris*, *Allium triquetrum* L., *Borago officinalis* L., and *Rumex acetosella* L., etc. Furthermore, other local plant foods are described.

Sardinia has the highest proportion of GFPs (257 taxa for ca. 2,100 in the vascular plant flora) in all the areas analyzed (table 5) [124]. The repertory of GFP recipes and local names is extremely diverse here, including the complex 'minestra delle 18 erbe selvatiche' at Barbagia [124–126]. Liquors are here traditionally made with white and blue fruits of *Myrtus communis* [S. Ríos, pers. commun.].

The prevalence of extreme longevity (on the island of Sardinia) or the healthy ageing (Crete) among the islanders of the Mediterranean resulted in considerable research interest in their genetic traits but also on their lifestyle and local food. An epidemiological study in all 233 centenarians living in Sardinia [127] showed that the prevalence of centenarians was 13.56 per 100,000, and the female/male ratio was approximately 2, instead of 5 found in other countries with reliable data [128, 129]. Passarino et al. [130] analyzed a sample of 40 centenarians and 116 young controls from Sardinia, using a set of new Y-chromosome binary markers. The results indicate that none of the seven lineages that account for >97% of the Y-chromosome diversity in Sardinia provide an advantage with respect to the extreme longevity. Therefore, the authors conclude that *there is no evidence for a genetic origin of longevity.* Their results also indicate that the Sardinian population had two main founder populations that have evolved in isolation for at least the last 5,000 years.

The three main Mediterranean islands (Corsica, Sicily and Sardinia) show evidence for different patterns of human peopling, with Corsica and Sicily closely associated with neighboring continental populations, while Sardinia shows a marked feature of isolation, with some possible ancient contact with the Iberian Peninsula. Approximately 60% of the Sicilian haplotypes are also prevalent in Southern Italy and Greece. Conversely, the Corsican sample had elevated levels of alternative haplotypes common in Northern Italy. Sardinia showed a haplotype ratio similar to that observed in Corsica, but with a remarkable difference in the presence of a lineage defined by marker M26, which approaches 35% in Sardinia but seems absent in Corsica. Although geographically adjacent, the data suggest different colonization histories and a minimal amount of recent gene flow between them. The results identify possible ancestral continental sources of the various island populations and underscore the influence of founder effect and genetic drift [131].

Conclusions

Ethnobotanical research has identified ca. 2,300 different plant and fungi taxa, which are gathered and consumed in the Mediterranean. Among these over 1,000 are only consumed in one single zone, therefore are strictly local. The percentage of local GFP taxa (present in <5 samples, table 5) is higher in the main centers of diversity at the periphery of the Mediterranean (Sahara, Alps, Caucasus, Canary Islands, the Levant). Islands (Sicily, Sardinia, Crete, Cyprus) also show a high proportion. Endemism of GFP taxa only accounts for a limited number of these 'ethnobotanical endemics' (only ca. 350 are endemic/endangered species). Are thus biological endemics of minimal medico-ethnobotanical importance? (See Moerman [263] for North America.) On the other hand, only a few taxa – 30 occurring in >20 samples – are consumed in most of the Mediterranean countries. Most have been analyzed in the Local Food-Nutraceuticals project [10]. These are missing in the Sahara and in low proportion in the Eastern Mediterranean samples. The ca. 800 GFP taxa that occur in more than the 5% of localities show a geographical pattern that permits one to recognize seven geographical groups. These groups show relationships with types of Mediterranean diet and could also be related with human genetic polymorphism through long-term co-evolution in a geographical mosaic pattern. Modulating this co-evolution are nutrients and secondary metabolites present in the GFPs.

The analysis by Simopoulos [132] of α-tocopherol, total phenols, antioxidant activity and antiradical power of Cretan GFPs which were studied as part of the Local Food-Nutraceuticals project, focus on the promising features as food of these scarcely known species. Wild edible plants contain altogether nutritional (vitamins, fatty acids, etc.), medicinal substances and toxic factors (oxalic acid, nitrate, erucic acid) [133–135]. Tumino et al. [136] have found the GFPs traditionally eaten in Ragusa Province (Italy) to be an important source of dietary antioxidants, as did Pieroni et al. [137] with those consumed by ethnic Albanians in Southern Italy.

The results of the Local Food project [10] are an excellent example of the complex interactions between GFPs and human metabolism.

The integration of genetic and cultural diversity in terms of food and food patterns is still a major endeavor. As pointed out many times, *there is an inextricable link between cultural and biological diversity*. This principle was first formulated at the 1st International Congress on Ethnobiology in Belem in the year 1988 (138: 208 and passim). Here we show the complexity of these links, which are crucial for understanding the interaction between humans and their environment and will require an integration specifically of genetic, ethnobiological and epidemiological research. More specifically, determining in detail the relationships between cultural and genetic features of the different ethnic groups around the Mediterranean and their food preferences can help to assess more clearly potential health benefits of the Mediterranean diets.

References

1 Hedrick UP: Sturtevant's Notes on Edible Plants. Geneva/NY, New York Agricultural Experimental Station, 1919.

2 Tanaka T: Tanaka's Cyclopaedia of Edible Plants of the World. Tokyo, Keigaku, 1976.

3 Uphof J: Dictionary of Economic Plants. Cramer, Lehre, 1968.

4 Rivera D, Obón C: La Guía de Incafo de las Plantas Útiles y Venenosas de la Península Ibérica y Baleares. Madrid, INCAFO, 1991.

5 Grossheim A: Rastitelnie Botatstva Kavkasa. Moscow, Moskovskoe Obtshestvo Islitatelei Prirodi, 1952.

6 Le Floc'h E: Contribution a une étude ethnobotanique de la flore tunisienne. Tunis, Imprimerie Officielle de la République Tunisienne, 1983.

7 Alcorn JB: The scope and aims of ethnobotany in a developing world; in Schultes RE, Von Reis S (eds): Ethnobotany – The Evolution of a Discipline. Portland, Dioscorides Press, 1997, pp 23–39.

8 Heinrich M: Ethnobotanik und Ethnopharmazie. Eine Einführung. Wissenschaftliche Verlagsgesellschaft, Stuttgart, 2001.

9 Rivera D, Obon C, Inocencio C, Heinrich M, Verde A, Fajardo J, Llorach R: The ethnobotanical study of local Mediterranean food plants as medicinal resources in Southern Spain. J Physiol Pharmacol 2005;56(suppl 1):97–114.

10 The Local Food-Nutraceutical Consortium: Understanding local Mediterranean diets: a multidisciplinary pharmacological and ethnobotanical approach. Pharmacol Res 2005;52:353–366.

11 Heinrich M, Leonti M, Nebel S, Peschel W: 'Local Food-Nutraceuticals': an example of a multidisciplinary research project on local knowledge. J Pharmacol Physiol 2005;56(suppl):5–22.

12 Cavalli-Sforza LL, Feldman M: The application of molecular genetic approaches to the study of human evolution. Nat Genet Suppl 2003;33:266–275.

13 Stinchcombe J, Rausher M: The evolution of tolerance to deer herbivory: modifications caused by the abundance of insect herbivores. Proc R Soc Lond B 2002;269:1241–1246.

14 Hollox E, Poulter M, Zvarik M, et al: Lactase haplotype diversity in the Old World. Am J Hum Genet 2001;68:160–172.
15 Ordovas J, Corella D: Nutritional genomics. Annu Rev Genomics Hum Genet 2004;5:71–118.
16 Heinrich M, Kufer J, Leonti M: Ethnobotanik und Pharmaziegeschichte – gemeinsame Herausforderungen und Aufgaben. Z Phytother 2005;26:54–60.
17 Leonti M, Sticher O, Heinrich M: Antiquity of medicinal plant usage in two Macro-Mayan ethnic groups. J Ethnopharmacol 2003;88:119–124.
18 Chikhi L, Nichols R, Barbujani G, Beaumont M: Y genetic data support the Neolithic demic diffusion model. Proc Natl Acad Sci USA 2002;99:11008–11013.
19 Richards M: The Neolithic invasion of Europe. Annu Rev Anthropol 2003;32:135–162.
20 Semino O, Passarino G, Oefner PJ, Lin AA, Arbuzova S, et al: The genetic legacy of Palaeolithic *Homo sapiens sapiens* in extant Europeans: a Y-chromosome perspective. Science 2000;290: 1155–1159.
21 Cavalli-Sforza LL: Genes, Peoples and Languages. London, Alien Lane, 2000.
22 Renfrew C: At the edge of knowability: towards a prehistory of languages. Cambridge Archaeol J 2000;10:7–34.
23 Tattersall I, Schwartz J: Hominids and hybrids: the place of Neanderthals in human evolution. Proc Natl Acad Sci USA 1999;96:7117–7119.
24 Bar-Yosef O: The upper Paleolithic revolution. Annu Rev Anthropol 2002;31:363–393.
25 Storz J, Payseur B, Nachman M: Genome scans of DNA variability in humans reveal evidence for selective sweeps outside of Africa. Mol Biol Evol 2004;21:1800–1811.
26 Etler D: *Homo erectus* in East Asia: human ancestor or evolutionary dead-end. Athena Rev 2004;4:37–50.
27 Harpending H, Rogers A: Genetic perspectives on human origins and differentiation. Annu Rev Genomics Hum Genet 2000;1:361–385.
28 Torroni A, Bandelt H, Macaulay V, et al: A signal, from human mtDNA, of postglacial recolonization in Europe. Am J Hum Genet 2001;69:844–852.
29 Achilli A, Rengo C, Magri CA, et al: The molecular dissection of mtDNA haplogroup H confirms that the Franco-Cantabrian glacial refuge was a major source for the European gene pool. Am J Hum Genet 2004;75:910–918.
30 Forster P: Ice Ages and the mitochondrial DNA chronology of human dispersals: a review. Phil Trans R Soc Lond B 2004;359:255–264.
31 Loogväli E, Roostalu U, Malyarchuk B, et al: Disuniting uniformity: a pied cladistic canvas of mtDNA haplogroup H in Eurasia. Mol Biol Evol 2004;21:2012–2021.
32 Richards M, Macaulay V, Hickey E, et al: Tracing European founder lineages in the Near Eastern mtDNA pool. Am J Hum Genet 2000;67:1251–1276.
33 Richards M, Macaulay V, Torroni A, Bandelt H: In search of geographical patterns in European mitochondrial DNA. Am J Hum Genet 2002;71:1168–1174.
34 Scozzari R, Cruciani F, Pangrazio A, et al: Human Y-chromosome variation in the Western Mediterranean area: implications for the peopling of the region. Hum Immunol 2001;62:871–884.
35 Ammerman A, Cavalli-Sforza LL: The Neolithic transition and the genetics of population in Europe. Princeton, Princeton University Press, 1984.
36 Cavalli-Sforza LL, Menozzi P, Piazza A: The History and Geography of Human Genes. Princeton, Princeton University Press, 1994.
37 Quintana-Murci L, Veitia R, Fellous M, Semino O, Poloni E: Genetic structure of Mediterranean populations revealed by Y-chromosome haplotype analysis. Am J Phys Anthropol 2003;121: 157–171.
38 Barbujani G, Bertorelle G: Genetics and the population history of Europe. Proc Natl Acad Sci USA 2001;98:22–25.
39 Zerjal T, Beckman L, Beckman G, et al: Geographical, linguistic, and cultural influences on genetic diversity: Y-chromosomal distribution in Northern European populations. Mol Biol Evol 2001;18:1077–1087.
40 Poloni E, Semino O, Passarino G, et al: Human genetic affinities for Y-chromosome P49a,f/Taql haplotypes show strong correspondence with linguistics. Am J Hum Genet 1997;61:1015–1035.

41 Cinnioglu C, King R, Kivisild T, et al: Excavating Y-chromosome haplotype strata in Anatolia. Hum Genet 2004;114:127–148.
42 Nettle D, Harris L: Genetic and linguistic affinities between human populations in Eurasia and West Africa. Hum Biol 2003;75:331–344.
43 Ethnologue, 2005: Languages of the world. http://www.ethnologue.com/
44 Alinei M, 2005: The Palaeolithic continuity theory on Indo-European origins: an introduction. http://www.continuitas.com/intro.pdf
45 McMahon R: Genes and Languages. Community Genet 2004;7:2–13.
46 Renfrew C: Archaeology and Language. Harmondsworth, Penguin, 1987.
47 Cavalli-Sforza LL, Minch E: Palaeolithic and Neolithic lineages in the European mitochondrial gene pool. Am J Hum Genet 1997;61:247–251.
48 Alinei M: European dialects: a window on the prehistory of Europe. Lingua Stile 2001;36:219–240.
49 Diamond J, Bellwood P: Farmers and their languages: the first expansions. Science 2003;300: 597–603.
50 Nasidze I, Stoneking M: Mitochondrial DNA variation and language replacements in the Caucasus. Proc R Soc Lond B 2001;268:1197–1206.
51 Conservation International, 2005: Biodiversity hotspots. Mediterranean basin. http://www.biodiversityhotspots.org
52 Davis SD, Heywood VH, Hamilton AC: Centres of Plant Diversity. Oxford, WWF, 1994.
53 Colledge S, Conolly J, Shennan S: Archaeobotanical evidence for the spread of farming in the Eastern Mediterranean. Curr Anthropol 2004;45(suppl):35–58.
54 Rivera D, Obón C: Palaeoethnobotany of compositae in Europe, North Africa and the Near East; in Caligari P, Hind D (eds): Compositae: Biology and Utilization. Kew, Royal Botanic Gardens, 1996, vol 2, pp 515–545.
55 Von der Hiort H: *Plantae esculentae patria*; in Linnaeus C (ed): Amoenitates Academicae. Stockholm, Laurent Salvii, 1756, vol 3, pp 74–99.
56 Von der Burg H: *Acetaria*; in Linnaeus C (ed): Amoenitates Academicae. Stockholm, Laurent Salvii, 1759, vol 4, pp 536–552.
57 Jerlin P: *Macellum olitorium*; in Linnaeus C (ed): Amoenitates Academicae. Leyden, Wetsten, 1764, vol 6, pp 116–131.
58 Salberg J: *Fructus esculenti*; in Linnaeus C (ed): Amoenitates Academicae. Leyden, Wetsten, 1764, vol 6, pp 342–364.
59 Keys A, Keys M: How to Eat Well and Stay Well: The Mediterranean Way. New York, Doubledays, 1959.
60 Hoffman W, 2003: Meet Monsieur Cholesterol. http://mbnet.umn. edu/hoff/hoff_ak.html
61 Keys A, Seven Countries: A Multivariate Analysis of Death and Coronary Heart Disease. Cambridge, Harvard University Press, 1980.
62 Keys A, et al: Coronary heart disease among Minnesota business and professional men followed 15 years. Circulation 1963;28:381–395.
63 Naska A: Mediterranean diet at present; in Barnat S (ed): International Conference on Health Benefits of Mediterranean Diet. Abstract Book. Aghia Pelagia, EGEA, 2003, p 21.
64 Trichopoulou A, et al: Diet and survival of elderly Greek: a link to the past. Am J Clin Nutr 1995;61:1346–1350.
65 Noah A, Truswell A: There are many Mediterranean diets. Asia Pacific J Clin Nutr 2001;10:2–9.
66 Simopoulos A: The Mediterranean diets: what is so special about the diet of Greece? The scientific evidence. J Nutr 2001;131:3065–3073.
67 Simopoulos A, Visioli F: Mediterranean Diets. Basel, Karger, 2001.
68 Gasimov G, Hashimova U, Abbasov A, Ismaylova A, 2002: Centenarians in Azerbaijan. Heredity as stored environment. http://www.azeri.org
69 Alakbarov F, 2000: Nutrition for longevity. Azerbaijan International Autumn 2000. http://www.globalaging.org/health/world/azerbarticle.htm
70 Anonymous, 2005: Cuisine of Azerbaijan. http://azerbaijan.tourism.az
71 Harmon N: The Mediterranean Diet Cookbook. New York, Bantam Books, 1994.
72 Rivera D, et al. (eds): Guía Etnobotánica de los Alimentos locales recolectados en la Provincia de Albacete. Albacete, Instituto de Estudios Albacetenses Don Juan Manuel, 2005.

73 Olver L, 1999: The food timeline. http://www.foodtimeline.org
74 Kiple K, Ornelas K: The Cambridge World History of Food. Cambridge, Cambridge University Press, 2000.
75 Hort AF: Theophrastus Enquiry into Plants. II. Cambridge, Harvard University Press, 1980.
76 Dalby A: Siren Feasts: A History of Food and Gastronomy in Greece. London, Routledge, 1996.
77 Savvides L: Edible wild plants of the Cyprus Flora. Nicosia, Savvides, 2000.
78 Lambraki M: Herbs, Greens, Fruit. The Key to the Mediterranean Diet. Heraklion, Lambraki, 2001.
79 Ertug F: An ethnobotanical study in Central Anatolia (Turkey). Econ Bot 2000;54:155–182.
80 Ozbucak T, Kutbay H, Akcin O: The contribution of wild edible plants to human nutrition in the Black Sea region of Turkey. 2006. http://www.siu.edu/~ebl/leaflets/tugba.htm
81 Ertug F: Wild edible plants of the Bodrum area (Mugla, Turkey). Turk J Bot 2004;28:161–174.
82 Dogan Y, Baslar S, Ay G, Mert H: The use of wild edible plants in Western and Central Anatolia (Turkey). Econ Bot 2004;58:684–690.
83 Huici A: Traducción española de un manuscrito anónimo del Siglo XIII sobre la cocina Hispano-Magribí. Madrid, Ayuntamiento de Valencia, 1966.
84 Crowfoot G, Baldensperger L: From Cedar to Hyssop. A Study in the Folklore of Plants in Palestine. London, Sheldon Press, 1932.
85 Carmona MD, Llorach R, Obón C, Rivera D: 'Zahraa', a Unani multicomponent herbal tea widely consumed in Syria: components of drug mixtures and alleged medicinal properties. J Ethnopharmacol 2005;102:344–350.
86 Tukan SK, Takruri HR, al-Eisawi DM: The use of wild edible plants in the Jordanian diet. Int J Food Sci Nutr 1998;49:225–235.
87 Machatschek M: Nahrhafte Landschaft. Vienna, Böhlau Verlag, 2003.
88 Machatschek M: Nahrhafte Landschaft. 2. Vienna, Böhlau Verlag, 2004.
89 Poletti A, von Niederhäusern-Terrani F: Erbe Selvatiche dal Ticino alla Valle d'Aosta. Pregassona, Fontana Edizioni, 2004.
90 Moroder M: Heilpflanzen in den Dolomiten. Bozen, Athesia, 2003.
91 Paume MC: Sauvages et comestibles. Herbes, fleurs et petites salades. Aix-en-Provence, Édisud, 2005.
92 Rivière-Sestier M: Remèdes populaires en Dauphiné. Grenoble, Presses Universitaires de Grenoble, 2000.
93 Andreotti G, Vernaccini S, Andreotti A: Guida Enogastronomica del Trentino. Trento, Artimedia, 2002.
94 Medvedev ZA: Caucasus and Altai longevity: a biological or social problem? Gerontologist 1974;14:381–387.
95 Nasidze I, Risch G, Robichaux M, et al: Alu insertion polymorphisms and the genetic structure of human populations from the Caucasus. Eur J Hum Genet 2001;9:267–272.
96 Nasidze I, Sarkisian T, Kerimov A: Testing hypotheses of language replacement in the Caucasus: evidence from the Y-chromosome. Hum Genet 2003;112:255–261.
97 Nasidze I, Quinque D, Dupanloup I, et al: Genetic evidence concerning the origins of South and North Ossetians. Ann Hum Genet 2004;68:588–599.
98 Nasidze I, Ling E, Quinque D, et al: Mitochondrial DNA and Y-chromosome variation in the Caucasus. Ann Hum Genet 2004;68:205–221.
99 Renfrew C: Archaeology, genetics, and linguistic diversity. Man 1992;27:445–478.
100 Jones VHS: Pliny Natural History Books 20–23. Cambridge, Harvard University Press, 1999.
101 Rackham H: Pliny Natural History Books 17–19. Cambridge, Harvard University Press, 1997.
102 Rackham H: Pliny Natural History Books 12–16. Cambridge, Harvard University Press, 2000.
103 Österman M: *Culina mutata*; in Linnaeus C (ed): Amoenitates Academicae. Stockholm, Laurent Salvii, 1760, vol 5, pp 120–132.
104 Asso IJ: Synopsis Stirpium Indigenarum Aragoniae. Marseilles 1779.
105 Cienfuegos B: Historia de las Plantas (MS now at the Biblioteca Nacional de España, Madrid), 1627.
106 Chabert A: De l'emploi populaire des plantes sauvages. Chambéry 1897.
107 Villars D: Catalogue des substances végétales, qui peuvent server à la nourriture de l'homme et qui se trouvent dans les départements de l'Isère, la Drôme et les Hautes Alpes, 1794.

108 Gaut JB: La salade champêtre. Rev Agric Forest Provence 1871:52–60.
109 Reguis M: Nomenclature franco-provençale des plantes qui croissent dans notre región. Mém Acad Sci d'Aix 1878;11:11–186.
110 Chauvet H, et al: Les salades sauvages. L'ensalada champanèla. Ecologistes de l'Euzière, 2003.
111 Bonet MA, Vallès J: Use of non-crop food vascular plants in Montseny biosphere reserve (Catalonia, Iberian Peninsula). Int J Food Sci Nutr 2002;53:225–248.
112 Brion M, Salas A, González-Neira A, Lareu M, Carracedo A: Insights into Iberian population origins through the construction of highly informative Y-chromosome haplotypes using biallelic markers, STRs, and the MSY1 minisatellite. Am J Phys Anthropol 2003;122:147–161.
113 González A, Brehm A, Pérez A, Maca-Meyer N, Flores C, Cabrera V: Mitochondrial DNA affinities at the Atlantic fringe of Europe. Am J Phys Anthropol 2003;20:391–404.
114 Comas D, Calafell F, Benchemsi N, et al: Alu insertion polymorphisms in NW Africa and the Iberian Peninsula: evidence for a strong genetic boundary through the Gibraltar Straits. Hum Genet 2000;107:312–319.
115 Bellakhdar J: La pharmacopée marocaine traditionelle. Paris, Ibis Press, 1997.
116 Gast M: Moissons du desert. Utilisation des resources naturelles au Sahara central. Paris, Ibis Press, 2000.
117 Nebel A, Landau E, Filon D, Oppenheim A, Fareman M: Genetic evidence for the expansion of Arabian tribes into the Southern Levant and North Africa. Am J Hum Genet 2002;70: 1594–1596.
118 Maca-Meyer N, González A, Pestano J, et al, 2003: Mitochondrial DNA transit between West Asia and North Africa inferred from U6 phylogeography. http://www.biomedcentral.com/1471–2156/4/15
119 Williams P: Giuseppe Pitre. Sicilian Folk Medicine. Lawrence, Coronado Press, 1971.
120 Picchi G, Pieroni A: Atlante dei prodotti tipici. Le Erbe. Roma, AGRA RAI-ERI, 2005.
121 Arcidiacono S, Pavone P, Salmeri C, 2002: Le Erbe commestibili dell'Etna. http://www.unict.it/dipartimenti/biologia_animale
122 Arcidiacono S, Pavone P, Salmeri C, 2005: Le piante Alimurgiche. http://www.dipbot.unict.it/alimurgiche
123 Simonpoli P: Arburi, Arbe, Arbigliule. Savoirs populaires sur les plantes de Corse. Ajaccio, Parc Naturel Regional de la Corse, 1993.
124 Atzei AD: Le piante nella tradizione popolare della Sardegna. Sassari, Carlo Delfino Editore, 2003.
125 Caria R: Le Piante Spontanee ad Alghero. Sassari, Editrice Democratica Sarda, 2001.
126 Congia P: Dizionario Botanico Sardo. Cagliari, Zonza Editori, 1998.
127 Deiana L, Ferrucci L, Pes GM, et al: The Sardinia Study of Extreme Longevity. Aging (Milan) 1999;11:142–149.
128 Koenig R: Sardinia's mysterious male methuselahs. Science 2001;291:2074–2076.
129 Koenig R: Demography: an Island of 'genetic parks'. Science 2001;291:2075.
130 Passarino G, Underhill P, Cavalli-Sforza LL, et al: Y-chromosome binary markers to study the high prevalence of males in Sardinian centenarians and the genetic structure of the Sardinian population. Hum Hered 2001;52:136–139.
131 Francalacci P, Morelli L, Underhill P, et al: Peopling of three Mediterranean islands (Corsica, Sardinia, and Sicily) inferred by Y-chromosome biallelic variability. Am J Phys Anthropol 2003;121:270–279.
132 Simopoulos A: Omega-3 fatty acids and antioxidants in edible wild plants. Biol Res 2004;37: 263–277.
133 Guil J, Rodríguez I, Torija E: Nutritional and toxic factors in selected wild edible plants. Plant Foods Hum Nutr 1997;51:99–107.
134 Guil J, Giménez I, Rodríguez M, Torija E: Nutritional composition of *Sonchus* species (*S. asper* L., *S. oleraceus* L. and *S. tenerrimus* L.). J Sci Food Agric 1998;76:628–632.
135 Guil J, Gimenez I, Torija M: Mineral nutrient composition of edible wild plants. J Food Compost Anal 1998;11:322–328.
136 Tumino R, Frasca G, Pellegrini N, Salvatore S, Krogh V, 2004: Antioxidant intake from traditional edible wild greens in the EPIC cohort of Ragusa (southeast Sicily). http://www.a-nutritional-supplements.com/conf04a38.htm

137 Pieroni A, Janiak V, Dürre, et al: In vitro antioxidant activity of non-cultivated vegetables of ethnic Albanians in Southern Italy. Phytother Res 2002;16:467–473.
138 Posey DA: Indigenous Knowledge and Ethics: A Darell Posey Reader. London, Routledge, 2004 [Posthumous edition by K. Plenderleith in Studies in Environmental Anthropology, vol 10].
139 Sykes B: The molecular genetics of European ancestry. Phil Trans R Soc B 1999;354:131–140.
140 Simoni L, Calafell F, Pettener D, Bertranpetit J, Barbujani G: Geographic patterns of mtDNA diversity in Europe. Am J Hum Genet 2000;66:262–278.
141 Torroni A, Bandelt H, D'Urbano L, et al: MtDNA analysis reveals a major Palaeolithic population expansion from southwestern to northeastern Europe. Am J Hum Genet 1998;62:1137–1152.
142 Underhill PA, Shen P, Lin AA, Jin L, Passarino G, et al: Y-chromosome sequence variation and the history of human populations. Nat Genet 2000;26:358–361.
143 Tardío J, Pascual H, Morales R: Alimentos silvestres de Madrid. Madrid, Comunidad de Madrid, 2002.
144 Rivera D, Obón C, Cano F, Robledo A: Introducción al mundo de las plantas medicinales en Murcia. Murcia, Ayuntamiento de Murcia, 1994.
145 Agelet A: Estudis d'Etnobotanica farmacèutica al Pallars; unpubl PhD thesis, Barcelona 1999.
146 Bauer AL: Raccogliere, Conservare, Cucinare, Funghi. Trento, Casa Editrice Publilux, 1992.
147 Blanco E, Cuadrado C: Etnobotánica en Extremadura. Estudio de la Calabria y la Siberia extremeñas. Alcoba de Los Montes, Emilio Blanco & CEP, 2000.
148 Blanco E: Diccionario de Etnobotánica Segoviana. Segovia, Ayuntamiento de Segovia, 1998.
149 Cervera M: Análisis comercial del sector de la seta silvestre en Cataluña. Distribución detallista en la ciudad de Lleida; Lic thesis, Lleida 1997.
150 Duran N, Morguí M, Sallés M: Plantes silvestres comestibles. Barcelona, Pòrtic, 2004.
151 Ferrández JV, Sanz JM: Las plantas en la medicina popular de la comarca de Monzón (Huesca). Huesca, Instituto de Estudios Altoaragoneses, 1993.
152 Giner JF: Estudio de las plantas comestibles silvestres de la Marina Alta (Alicante). Trabajo fin de carrera de Ingeniero Técnico Agrícola. Escuela Politécnica Superior de Orihuela. Orihuela, Universidad Miguel Hernández, 2004.
153 Guarrera PM: Food medicine and minor nourishment in the folk traditions of Central Italy (Marche, Abruzzo and Latium). Fitoterapia 2003;74:515–544.
154 Guerau G, Torres N: Nova aportació al coneixement de les plantes d'Eivissa I Formentera. Eivissa, Consell Insular, 1981.
155 Inocencio C, Rivera D, Obón C, Verde A, Heinrich M: Local food as a sustainable development issue in SE Spain. S Afr J Bot 2003;69:251.
156 Lastra JJ: Etnobotánica en el Parque Nacional de Picos de Europa. Madrid, Ministerio de Medio Ambiente, 2003.
157 Márquez R: Estudio de las plantas comestibles silvestres del Bajo Segura. Trabajo fin de carrera de Ingeniero Técnico Agrícola. Escuela Politécnica Superior de Orihuela. Orihuela, Universidad Miguel Hernández, 2003.
158 Martínez MJ, González MR, Molero J: Investigaciones Etnobotánicas en el Parque Natural del Cabo de Gata-Nijar (Almería). Almería, Sociedad Almeriense de Historia Natural, 1997.
159 Mateo G, Crespo B: Manual para la determinación de la flora valenciana. Valencia, Moliner, 2002.
160 Mesa S: Estudio etnobotanico agroecológico de la comarca de la Sierra de Mágina (Jaén); unpubl PhD thesis, Madrid 1996.
161 Morales R, Tardío F, Pascual H, Laca L, Alarcón R, García P, 2004: Etnobotánica Alimentaria. http://www.terra.es/personal3/etnobotanica/home.htm
162 Muntané J: Herbes, Remeis I Creences de Cerdanya. Ripoll, Institut d'Estudis Ceretans, 1994.
163 Nicolás C: Estudio de las plantas comestibles silvestres del municipio de Murcia. Trabajo fin de carrera de Ingeniero Agrónomo. Escuela Politécnica Superior de Orihuela. Orihuela, Universidad Miguel Hernández, 2005.
164 Oltra JE: Fer Herbes a Quatretonda. Alicante, Collectiu Cultural Dorresment, 1999.
165 Palau P: Plantes Medicinals Baleariques. Palma de Mallorca, Editorial Moll, 1993.
166 Pellicer J: Costumari Botànic. Valencia, Edicions del Bullent, 2001.
167 Pignatti S: Flora d'Italia. Bologna, Edagricole, 1982.
168 Pieroni A: Evaluation of the cultural significance of wild food botanicals traditionally consumed in northwestern Tuscany, Italy. J Ethnobiol 2001;21:89–104.

169 Pieroni A, Nebel S, Quave C, et al: Ethnopharmacology of liakra: traditional weedy vegetables of the Arbëreshë of the Vulture area in Southern Italy. J Ethnopharmacol 2002;81:165–185.
170 Vangjeli J: Udhëheqës Fushor I Florës Së Shqipërisë. Tirana, Academy of Sciences, 2003.
171 Arnold N: Contribution a la connaissance ethnobotanique et medicinale de la flore de Chipre; PhD thesis, Paris 1985.
172 Polunin O: Flowers of Greece and The Balkans. Oxford, Oxford University Press, 1980.
173 Trichopoulou A, Vasilopoulou E, Hollman, P, et al: Nutritional composition and flavonoid content of edible wild greens and green pies: a potential rich source of antioxidant nutrients in the Mediterranean diet. Food Chem 2000;70:319–323.
174 Hirsch EG, Löw I, 2005: Plants in the Bible. http://www.jewish encyclopedia.com
175 Moldenke HN, Moldenke AL: Plants of the Bible. London, Ronald Press, 1952.
176 Plitmann U, Heyn C, Danin A, Shmida A: Pictorial Flora of Israel. Jerusalem, Massada, 1982.
177 Mifsud S, 2005: Malta wild plants. http://www.maltawildplants.com
178 Harlan SM, Sell PD, Wolseley H: A Flora of the Maltese Islands. Msida, Malta University Press, 1977.
179 Baytop T: Türkiyede Bitkiler ile Tedavi. Istanbul, Istanbul University, 1984.
180 Özgen U, Kaya Y, Coskun M: Ethnobotanical studies in the villages of the district of Ilica (Province Erzurum), Turkey. Econ Bot 2004;58:691–696.
181 Öztürk M, Özçelik H: Dogu Anadolu'nun Faydali Bitkileri. Ankara, Siskav, 1991.
182 Simsek I, Aytekin F, Yesilada E, Yildirimli S: An ethnobotanical survey of the Beypazari, Ayas and Güdül district towns of Ankara Province (Turkey). Econ Bot 2004;58:705–720.
183 Afyon A, Yagiz D, Konuk M: Macrofungi of Sinop Province. Turk J Bot 2004;28:351–360.
184 Bellakhdar J: Médecine traditionnelle et toxicologie ouest-saharienne. Rabat, Éditions Techniques Nord-Africaines, 1978.
185 Beloued A: Plantes médicinales d'Algérie. Algiers, Office des Publications Universitaires, 1998.
186 Benchelah A, Bouzane H, Maka M, Ouahès C: Fleurs du Sahara. Voyage ethnobotanique avec les Touaregs du Tassili. Paris, Ibi Press, 2000.
187 Foureau F: Essai de catalogue des noms Arabes et Berbères de quelques plantes, arbustes et arbres algériens et sahariens. Paris, Challamel, 1896.
188 Ghazanfar SA: Arabian Medicinal Plants. Boca Raton, CRC Press, 1994.
189 Guest E: Notes on Plants and Plant Products with their Colloquial Names in Iraq. Bull No 27. Baghdad, Department of Agriculture, 1933.
190 Hooper D, Field H: Useful Plants and Drugs of Iran and Iraq. Botanical Series Field Museum of Natural History 1937;9:73–241.
191 Miller AG, Morris M: Plants of Dhofa, the Southern Region of Oman. Traditional, Economic and Medicinal Uses. Government of Oman, 1988.
192 Perrot E, Gentil L: Sur les productions végétales du Maroc. Paris, Office National des Matiéres Premiéres Végétales, 1921.
193 Post GE, Dinsmore JE: Flora of Syria, Palestine and Sinai. Beirut, American University of Beirut, 1932.
194 Quezel P, Santa S: Nouvelle Flore de l'Algérie. Paris, Editions du CNRS, 1963.
195 Schweinfurth G: Arabische Pflanzennamen aus Ägypten, Algerien und Jemen. Berlin, Dietrich Reimer, 1912.
196 Shuaib L: Wildflowers of Kuwait. London, Starey International, 1995.
197 Lacoizqueta JM: Diccionario de los nombres eúskaros de las plantas. Pamplona, Imprenta Provincial, 1888.
198 García-Bona LM: Plantas comestibles. Guia de plantas y setas comestibles de desarrollo espontáneo en Navarra. Pamplona, Gobierno de Navarra, 1994.
199 López G: Botánica popular alavesa. Vitoria, Diputación Foral de Alava, 1975.
200 Boa E: Hongos silvestres comestibles. Perspectiva global de su uso e importancia para la población (Productos Forestales no madereros 17). Rome, FAO, 2005.
201 Guinea E: Catalogo razonado de las plantas del Sahara español. Anal Jard Bot Madrid 1948;8: 357–442.
202 Boulos L: Egyptian desert plants with promising economic potential. Arab Gulf J Sci Res 1989;7:91–108.

203 Goodman SM, Hobbs J: The ethnobotany of the Egyptian Eastern Desert: a comparison of common plant usage between two culturally distinct Bedouin groups. J Ethnopharmacol 1988;23:73–89.
204 Heneidy SZ, Bidak LM: Potential uses of plant species of the coastal Mediterranean region, Egypt. Pak J Biol Sci 2004;7:1010–1023.
205 Osborn DJ: Notes on medicinal and other uses of plants in Egypt. Econ Bot 1968;22:165–177.
206 García M: El bosque de laurisilva en la economía guanche. Santa Cruz de Tenerife, Aula de Cultura de Tenerife, 1989.
207 Pereira E: Ilhas de Zargo. Funchal, Editora, 1967.
208 Perez PL, Hernandez CE: Plantas medicinales o útiles en la Flora Canaria. La Laguna, Francisco Lemus, 1999.
209 Porto da Cruz V do: A Flora Madeirense na Medicina Popular. Broteria 1935;4:35–46, 71–78, 140–154.
210 Porto da Cruz V do: A flora madeirenses na medicina popular e na industria. Publicaçâo da Liga para Proteçâo da Natureza 1950;4:1–17.
211 Vieira R: Flora de Madeira o Interesse das Plantas endémicas Macaronésicas. Lisboa, Serviço Nacional de Parques, Reservas e Conservaçâo de Natureza, 1992.
212 Della A: Cyprus: Country report to the International Conference and Programme on Plant Genetic Resources. Rome, FAO, 1996.
213 Baser KHC: Current knowledge on the wild food and non-food plants of Turkey. Cah Options Méditer 1997;23:129–159.
214 Duplessy B, Duc B: Le Livre de la Truffe. Aix-en-Provence, Edisud, 1998.
215 Juen W, Wieser J, Böhn T: Das Tiroler Schnaps Buch. Innsbruck, Tyrolia, 2002.
216 Poletti A: Fiori e piante medicinali. Quart/Aosta, Musumeci Editore, 2004.
217 Traverso M: Funghi delle nostre Alpi. Chiari, Nordpress, 2004.
218 Vernaccini S: Piante e Erbe nelle Dolomiti. Trento, Artimedia, 2003.
219 Heywood V: Medicinal, Culinary and Aromatic Plants in the Near East. Proceedings of the International Expert Meeting. Rome, FAO, 1999.
220 Chilton L, Turland N, 2004: Flora of Crete: Supplement II, Additions 1997–2004. http://www.marengowalks.com/fcs.html
221 Maurizio G, Paoletti A, Dreon L, Lorenzoni G: Pistic, traditional food from Western Friuli, NE Italy. Econ Bot 1995;49:26–30.
222 Verde A: Estudio etnofarmacológico de tres áreas de montaña de Castilla-La Mancha; PhD thesis, Murcia 2002.
223 Gamisans J, Marzocchi J: La Flore endémique de la Corse. Aix-en-Provence, Edisud, 2003.
224 Torres M: Antropologia d'Eivissa I Formentera. Herbes, pastors, ses matances. Eivissa, Editorial Mediterrània, 1999.
225 Solé C: Fruits silvestres del Pirineu receptari. Tremp, Garsineu Edicions, 1996.
226 Pellicer J: Costumari Botànic. 2. Valencia, Edicions del Bullent, 2000.
227 Pellicer J: Costumari Botànic. 3. Valencia, Edicions del Bullent, 2004.
228 López G, López N, 2004: Flora de Madrid. http://www.rjb.csic.es/Biodiversidad/intro.html
229 Conca A, Oltra JE: Plantas medicinales y comestibles. Onteniente, Caixa Ontinyent, 2005.
230 Pieroni A, Nebel S, Franco R, Heinrich M: Food for two seasons: culinary uses of non-cultivated local vegetables and mushrooms in a South Italian village. Int J Food Sci Nutr 2005;56:245–272.
231 Nebel S: Ta chòrta. Piante comestibili tradizionali di Galliziano. London, School of Pharmacy, University of London, 2005.
232 Gredilla AF: Apuntes para la Corografía Botánica Vasco-Navarra. Barcelona, Atlas Geográfico, 1913.
233 Vazquez G: Plantas medicinales en el País Vasco. San Sebastián, Txertoa, 1995.
234 Aramburu N, Aramburu A: Estudio Etnobotánico en Usurbil (Guipuzcoa; País Vasco). Murcia, Universidad de Murcia, 2005 (unpubl).
235 Pieroni A: Gathered wild food plants in the upper valley of the Serchio River (Garfagnana), Central Italy. Econ Bot 1999;53:327–341.
236 Aizpuru I: Claves ilustradas de la Flora del País Vasco y territorios limítrofes. Vitoria, Eusko Jaurlaritza, 1999.
237 Pieroni A: Medicinal plants and food medicines in the folk traditions of the upper Lucca Province, Italy. J Ethnopharmacol 2000;70:235–273.

238 Pieroni A, Dibra B, Grishaj G, Grishaj I, Maçai S: Traditional phytotherapy of the Albanians of Lepushe, Northern Albanian Alps. Fitoterapia 2005;76:379–399.
239 Pieroni A, Giusti M, Münz H, Lenzarini C, Turkovic G, Turkovic A: Ethnobotanical knowledge of the Istro-Romanians of Zejane in Croatia. Fitoterapia 2003;74:710–719.
240 Fernández C, Guzmán A, Fernández AM, Camacho AM: Plantas medicinales y útiles en la Península Ibérica. Jaén, Herbario Jaén, 1999.
241 Figueiredo JJ: Flora Pharmaceutica e alimentar Portuguesa. Lisboa, Academia R das Sciencias, 1825.
242 Lev E, Amar Z: Ethnopharmacological survey of traditional drugs sold in the Kingdom of Jordan. J Ethnopharmacol 2002;82:131–145.
243 Alakbarov F: Medicinal plants used in medieval Azerbaijan phytotherapy. J Herb Pharmacother 2001;1:35–49.
244 Kasparek M, Grimm U: European trade in Turkish Salep with special reference to Germany. Econ Bot 1999;53:396–406.
245 Iatrou G, Kokkalou E: Rarity, conservation, importance and ethnopharmacological knowledge of the Greek Flora. Options Méditer 1997;23:65–74.
246 Malamas M, Marselos M: The tradition of medicinal plants in Zagori, Epirus (northwestern Greece). J Ethnopharmacol 1992;37:197–203.
247 Huxley A, Taylor W: Flowers of Greece. London, Chatto & Windus, 1984.
248 Baumann H: Greek Wild Flowers and Plant Lore in Ancient Greece. London, Herbert Press, 1996.
249 Barkoudah Y: Country profile: Syrian Arab Republic (Syria). Cah Options Méditer 1998;38: 275–294.
250 Dabbour I, Takruri H: Protein quality of four types of edible mushrooms found in Jordan. Plant Food Hum Nutr 2002;57:1–11.
251 Lancaster W, Lancaster F: People, Land and Water in the Arab Middle East. Amsterdam, Harwood Academic Publishers, 1999.
252 Oliveira R: Medicina pelas plantas. Lisboa, Livraria Progresso Editora, 1952.
253 Said Q, Khalil K, Fuldr S, Azaizeh H: Ethnopharmacological survey of medicinal herbs in Israel, the Golan Heights and the West Bank Region. J Ethnopharmacol 2002;83:251–265.
254 Lambraki M: Ta Xorta. Athens, Ellenika Grammata, 1997.
255 Rubin R: The Melagria: on anchorites and edible roots in Judaean Desert. Liber Annuus 2002;52: 347–352.
256 Anonymous, 2005: In Crete. Horta. http://www.in-crete.co.uk/horta.htm
257 Anonymous, 2005: In Crete. Herbs. http://www.in-crete.co.uk/herbs.htm
258 Gabrielian E, Zohary D: Wild relatives of food crops native to Armenia and Nakhichevan. Flora Mediter 2004;14:5–80.
259 Lorenzo ML, Ríos S: Etnobotánica de los minxos: tradición culinaria de los valles del Guadalest y el Algar (Marina Baixa, Alicante). Alicante, Universidad de Alicante, 2005 (unpubl DEA report).
260 Martínez V, Ríos S: Lesser known herbal spirits of the Valencia and Alicante Provinces (ES Spain); in Pascual MJ, et al. (eds): Industrial Crops and Rural Development. Murcia, AAIC, 2005, pp 417–425.
261 Al-Eiswi D, Takruri H: A checklist of wild edible plants in Jordan. Arab Gulf J Sci Res Agric Biol Sci 1989;B7:79–102.
262 Takruri H, Humeid M: Nitrate levels in edible wild herbs and vegetables common in Jordan. Nutr Health 1988;8:89–98.
263 Moerman DE: Heilpflanzen aus Nordamerika. Z Phytother 1997;18:20–33.
264 The Y Chromosome Consortium: A nomenclature system for the tree of human Y-chromosomal binary haplogroups. Genome Res 2002;12:339–348.

Diego Rivera
Departamento de Biología Vegetal, Facultad de Biología
Universidad de Murcia, E-30100 Murcia (Spain)
Tel. +34 968 36 49 94, Fax +34 968 36 39 63, E-Mail drivera@um.es

Heinrich M, Müller WE, Galli C (eds): Local Mediterranean Food Plants and Nutraceuticals. Forum Nutr. Basel, Karger, 2006, vol 59, pp 75–85

Disseminating Knowledge about 'Local Food Plants' and 'Local Plant Foods'

Diego Rivera[a], Michael Heinrich[c], Concepción Obón[b], Cristina Inocencio[a], Sabine Nebel[c], Alonso Verde[a], José Fajardo[a]

[a]Departamento de Biología Vegetal, Facultad de Biología, Universidad de Murcia, Murcia, [b]Departamento de Biología Aplicada, EPSO, Universidad Miguel Hernández, Orihuela, Alicante, Spain; [c]Centre for Pharmacognosy and Phytotherapy, The School of Pharmacy, University of London, London, UK

Abstract

Ethnobotanical approaches to the study of Mediterranean food plants offer novel ways for analyzing and preserving traditional knowledge and agrobiodiversity in the Mediterranean area. This article highlights our strategy to increase the awareness within traditional knowledge systems and encourage the continuous evolution of it, avoiding the loss of substantial parts of the local cultural and biological diversity. The strategy is part of a broader stream of thought, which does attempt to disseminate information locally in a multitude of ways, e.g. through a range of publications in rural or urban zones, to people with or without formal education, to children or the elderly. This article is a very personal account of the experience of the authors, but there is an urgent need to assess the impact of such activities on a broader level, and, also, to reassess the impact researchers have on the communities. Our clear impression in all field sites has been that the simple fact that such traditional knowledge systems are the focus of scientific investigation are an essential element of giving renewed sociocultural value to such knowledge and that activities like the ones described here are of great interest to the communities we worked in.

Knowledge emerges out of a complex process involving social, situational, cultural and institutional factors. The process takes place on the basis of existing conceptual frameworks and procedures and is affected by skills, orientations, experiences, interests, resources and patterns of social interaction. Moreover, knowledge is constructive in the sense that it is the result of a great number of decisions and selective incorporations of previous ideas, beliefs and images, but at the same time destructive of other possible frames of conceptualization and understanding [1].

Local knowledge – knowledge that is unique to a given culture or society – seemingly contrasts with the scientific and technical international knowledge system generated by universities, research institutions and private firms. Local knowledge is the basis for local-level decision-making in agriculture, health care, food preparation, education, natural-resource management, and a host of other activities in rural communities [2]. In real life though, the difference between traditional and scientific knowledge is not that great. Freeman [3] argues that both types of knowledge rest on the systematic gathering of empirical observations. The main difference lies in the methods used for collection and analysis of data. Scientific knowledge needs a wide range of methodical observations to establish a model of a situation, for instance to estimate the development of a certain stock of animals within an ecosystem. Before a biologist can come to a conclusion about the development of the stock, he must collect great amounts of quantitative data over some time. A local fisherman, who is familiar with the area, will react spontaneously to observations that deviate from the usual pattern. He will be observant to qualitative changes, signs that indicate that something unusual is happening. He will interpret such signs within the context of his experience and traditional knowledge (TK), and discuss his interpretations with fellow fishermen and neighbors.

There is consensus amongst scientists using various terms that local knowledge: (i) is linked to a specific place, culture or society; (ii) is dynamic in nature; (iii) belongs to groups of people who live in close contact with natural systems, and (iv) contrasts with 'modern' or 'Western formal scientific knowledge' [1]. Most traditional knowledge systems (TKS) are oral-based, often revealed through stories and legends.

For these reasons, it is often difficult to transmit ideas and concepts to those who do not share the language tradition and cultural experience. When language is threatened or diminished the cultural transmission of TK is jeopardized. Local food plants and particularly gathered food plants (GFPs) [4] constitute a good practical example of TK. People do not eat all 'edible' plants available in their environments but only a small part of this flora. What makes the difference is the cultural decision that is behind each repertory of GFPs.

There are many different factors that determine the choice of a precise plant as a food: abundance, availability, cultural preferences, processing technologies, ability to collect in the optimal period and, last but not least, genetic features of the consumers (e.g. presence of detoxifying enzymes) that allow the safe consumption of the plant.

Therefore, in every society, people select a few or some species from the many potentially 'edible' ones. While abundance may be one of the reasons for selecting a plant as a local food, it is not sufficient and in fact in many regions, local selection has decided against specific abundant species. Sometimes they are only used as fodder, or included in the restrictive group of 'famine food', or simply neglected. Ethnobotany has demonstrated that this selective local profile of food is found at different scales (local to regional) being part of TKS [5].

There have been many attempts to disseminate results of research or local development projects, which were headed by academics, church workers, members of NGOs and in some countries by state officials (e.g. by institutes involved in indigenous affairs). Such approaches take a variety of forms and include for example:

- Books or booklets in the local language written specifically for the communities [e.g. 6–8].
- Electronic media.
- Exhibitions and activities in (local) museums.
- The creation of small community botanical gardens.
- Creating rural dispensaries of herbal products or small-scale industries producing such products.

While this list is not comprehensive, it highlights the diversity of approaches used and also that so far this approach has not been critically reviewed and assessed from an ethnobotanical or ethnopharmaceutical perspective. While in the following we do not attempt such a comprehensive review, we highlight the possibilities of disseminating local knowledge using one example – an European Union-funded project on 'Local Food-Nutraceuticals'.

The first prerequisite for disseminating knowledge is a substantial body of knowledge, based on both scientific and local traditional data that ought to be disseminated. It is fundamental to distinguish the activities within the local TKS and those outside of it. Ethnobotanical approaches to the study of Mediterranean food plants offer novel ways for analyzing and preserving TK and agrobiodiversity in the Mediterranean area. Ethnobotany can help to determine precisely what plants are actually eaten by each ethnic group in a determined geographical and cultural context, and is therefore a basic tool for a consistent dissemination of knowledge within the different frameworks of local TKS.

This was the purpose of the Local Mediterranean Food Plants Project (2001–2004) with the financial support of the European Union (QLK1-2001-00173). Nearly 150 GFP taxa from Greece, Italy and Spain were investigated in order to assess their ethnobotanical features as well as their biological activities [4, 9].

Dissemination Framework

Dissemination activities have been (and will continue to be in the future) an essential element of this project, which has been described in more detail in other contributions to this volume (see Heinrich et al., this volume) and such activities have been addressed to a variety of groups:

- Inhabitants of rural areas especially in the regions where field studies have been conducted. These were always in collaboration with local associations of housewives, ecologists or others. Such specific courses in ethnobotany have to be eminently practical, where they are carried out by several visits to the field with the participants (all of them experts of the place and therefore potential informants).
- Exhibitions addressed to a wider audience in the cities and networks of urban communities of the region focusing on various ethnobotanical and ethnomycological topics. These were carried out in collaboration with the Spanish (Albacete) Open University (Universidad Popular de Albacete), the Institute of Estudios Albacetenses, Caja Murcia, the City Council of Murcia or the Mycological Society of Albacete. A remarkable example is the community fair of Albacete (September 2000), visited by more than 200,000 people. Associated to these exhibitions are books on plants and folk medicine [10], aromatic plants [11] and ethnobotany of Albacete [12].
- Workshops with the so-called Third Age, in collaboration with centers of Attention to the Third Age, of the city councils or the provinces (Diputaciones).
- Talks addressed to a wide public audience on specific topics of ethnobotany or brief courses of medicinal plants or general ethnobotany.
- Pupils and students currently undertaking formal education at primary and secondary level. With these groups we put into practice many elements of the ethnobotanical knowledge we had recorded. Specifically the goal was to promote the transmission of TK within the families, through activities that require the participation of the elder members in each family.

Overall, these activities are also important to get new contacts, for example to people with a minimum interest in the topic but usually predisposed to collaborate. On occasion, we have met with the contradiction of giving a 'counter-information' when using names of plants or uses, unknown or different to those that were part of the TK in the locality. This occurs often because substantial differences are found at very short distance – a few kilometers. However, ethnobotanical research is always a two-way process and it would be impossible and not desirable to avoid it.

Examples of Dissemination Activities in Southeastern Spain

All of the authors who also are members of the educational community, work in the dissemination of ethnobotanical and related knowledge each one in different settings. In the field of the non-formal education, the open universities (Universidades Populares) offer interesting possibilities to disseminate local knowledge. Particularly in case of the Popular University of Albacete, introductory courses offer unique opportunities for interactive exchanges of TK. In case of vernacular names of plants, those collected in the area were disseminated, avoiding the bibliographical ones. Part of the programme is a 2-year course on 'Folk Botany' introducing the different aspects of the ethnobotany of Albacete, and it also includes practical activities and trips as well as other elements of applied botany. The sequence of classes is based on the calendar of the nature, taking advantage of the best moments for each topic. The different didactic units are conducted at the time points when these activities take or took place traditionally. Women are in the majority (ca. 80%), this proportion is inversed in the courses on 'Fungi'. Participants are usually people with considerable formal education, with a keen interest in cultural matters and considerable free time. The average age is around 45 years. In general, they are very actively involved in the dynamics of the classroom.

Within the program 'University of the Experience', dedicated to retired people, an optional course on 'Useful Plants' is offered. The course, with 30 sessions, covers different traditional uses of plants. Here potential informants (representative of local TKS) usually also participate. The average age is about 70 (65–88) years. The profile is distributed equally between both sexes, with a particular interest from teachers and retired professors.

In different other short courses, we also use our field and TK data:

- 'Mushrooms' focusing on all aspects of mushroom biology and use, including local names and local uses.
- 'Uses of the Aromatic Plants' including practical activities related with traditional remedies.
- 'Flora of Albacete' focusing on relevant plant species in the TKS around the city of Albacete.

Sánchez et al. [13] provide an excellent and quite early example of the results of enhancing the communication about TK within the family. Verde and Fajardo [7, 14, 15] and Verde et al. [16] adapted TK to the curriculum needs and the educational plans in subjects such as Nature, Physics & Chemistry, and Social Sciences. Specifically these are now being used as part of the curriculum of secondary education in the Community of Castilla-La Mancha.

The popularization of printed works and multimedia took us, on the part of different Center of Professors and Resources of Castilla-La Mancha and Murcia, to impart formation courses relating to this topic (TKS and Ethnobotany).

Since 1986 a course in Ethnobotany for undergraduate students of Biology and Environmental Sciences has been offered at Murcia University. Some 1,000 students attended this course in the ca. 20 years it is running. A similar course for students of Agronomy has been given in the University Miguel Hernandez (Orihuela) since 1998. Specific materials were published by Rivera and Obón [17]. The target of both courses is to introduce students to the basic methodology and concepts of Ethnobotany and to promote the transmission of TK. The practicals include a research project in rural communities that led the students to discover for themselves the richness of TKS in the different zones. Students worked mainly in the area of southeastern Spain (Murcia, Almería, Alicante, Albacete) but also developed projects in other areas of Spain (Cáceres, Guipuzcoa, Ibiza, Guadalajara, Castellón, Lugo, etc.). Special attention was recently paid to the ethnobotanical study of immigrants from Ecuador, Morocco and Algeria, that now account for over 20% of population in many towns of Murcia and Alicante.

In the dissemination of knowledge, different technical resources have been used: multimedia, printed, electronic, etc. The approach depended on the circumstances, resource availability and targeted public:

- Multimedia: water resources [18], plants [19], sustainable agriculture [20].
- Educational material [7, 14–16, 21].
- Booklets [7].
- Books in Spanish: [10–12, 22, 23].

The Dissemination Activities Associated with the Local Food-Nutraceuticals Project

As part of the project 'Local Food-Nutraceuticals', the dissemination of local food knowledge in the region of origin has been a crucial part of the project and was addressed in a variety of ways (see Heinrich et al., this volume). The activities are largely based on the knowledge gained during the project [24]. We developed a strategy of dissemination at different levels including printed materials, multimedia curriculum materials and exhibitions.

Specifically, part of the project [16] developed curriculum materials for secondary education entitled 'La alimentación en Castilla-La Mancha: de la escasez al desperdicio (el valor de los alimentos locales y su utilización sostenible)' (fig. 1). This work targets a population of ca. 250,000 children and was presented to the *Certamen de Materiales Curriculares de la Comunidad Autónoma de Castilla-La Mancha* in September 2004 and it received the second prize. This book was published as a CD-ROM in 2004 (600 copies). It includes puzzles with pictures of local foods and local food plants with introductory texts. The CD is accompanied by an 8-page booklet introducing the book and the project 'Local Food-Nutraceuticals'. The distribution to all secondary education centers in Castilla-La Mancha has been agreed with the

Fig. 1. Cover page of the multimedia curricular materials developed specifically for secondary education, as part of the project.

Regional Minister of Education and its public presentation in Toledo, the capital city of Castilla-La Mancha. The book contains five didactic units:

- 'Evolución histórica de los alimentos' (Historical development of food).
- 'La alimentación tradicional Castellano-Manchega. La dieta mediterránea' (The traditional food in Castilla-La Mancha. The Mediterranean diet).
- 'De la escasez al desperdicio' (From scarcity to excess).
- 'Alimentación, nutrición y salud' (Food, nutrition and health).
- 'Procesado, conservación y etiquetado de los alimentos' (Processing, conservation and labeling of food).

Each unit includes objectives, contents, activity cards, and a bibliography.

Another book – largely in Spanish but also bilingual (Spanish-English) in certain parts such as Recipes and Pharmacology – entitled: 'Guía etnobotánica de los alimentos locales recolectados en la provincia de Albacete', is addressed to the general public [25]. The contributors to the book are members of the

Fig. 2. Posters exhibited in the Ninth Mycological Week of Albacete ('IX Semana Micologica', November 8–11, 2004) explaining the Local Food Project and the main results, focusing on mushrooms as local foods.

different partner of the consortium 'Local Food-Nutraceuticals'. The book consists of two main parts: (1) Ethnobotany, with an introduction, methodology, collection and conservation, guide to plant taxa (145 species), local traditional recipes (ca. 200 recipes and variants) (bilingual in English/Spanish, translation), and (2) Phytochemical and Pharmacological Aspects (contributions in English by different authors from the groups involved) coordinated by M. Leonti and M. Heinrich. With Spanish translation by C. Inocencio and D. Rivera.

Other dissemination activities included:

- Participation in the regional activities focusing on the popularization of science (European Science and Technology Week – Semana de la Ciencia y la Tecnología 2003 and 2004, Murcia, Spain, November 2003 and 2004). Posters, multimedia presentations and an open code computer program (Linux operative system) were used to address primary and secondary school children. These included puzzles about locally GFPs and local dishes studied in project 'Local Food-Nutraceuticals'. The puzzles were also included in the Murcia University distribution called Caldum 2004. Some 400 CDs were distributed to the people attending the Fair.
- Participation in the Ninth Mycological Week of Albacete ('IX Semana Micologica', November 8–11, 2004, with seven posters explaining the project and the main results focusing on mushrooms as local foods (fig. 2).
- Participation in the Ninth Nutrition and Food Fair in Albacete ('IX Feria de Alimentación y Gastronomia' 'ALIMENTA', November 19–21, 2004) attended by over 20,000 persons with a series of posters.

- Cooperation with the '¿Quieres recuperar los sabores de antaño', a campaign organized by the NGO Asociación de Naturalistas del Sureste and Consejería de Medio Ambiente y Ordenación del Territorio of Murcia focusing on the local plant food cultivars and the interest of conserving this diversity.

Similarly, a book *Ta chòrta* on results in a 'Graecanic' (Greek)-speaking community in Italy was published by Sabine Nebel [26], which highlights the use of food plants in a small community in Reggio di Calabria undergoing rapid cultural change. Again the specific goal was to disseminate the outcomes of the research in this region. *Ta chòrta* roughly corresponds to what the modern nutritional science nowadays calls green leafy vegetables. During the 8th century BC, parts of Southern Italy, as we know it today, were under Greek influence, and known as *Magna Graecia* [27]. The Greek influence continued over centuries until the end of the Byzantine Empire in 1453 AC, also called the Eastern Roman Empire. During the fieldwork, 48 different wild gathered food species were identified. Most of the recorded plant species are commonly in use in Southern Italy and the Mediterranean region. The book is currently (late 2005) distributed in the community and neighboring regions.

Conclusions

This article highlights our strategy to increase the awareness within TKS and encourage the continuous evolution of it, avoiding the loss of substantial parts of the local cultural and biological diversity. The strategy is part of a broader stream of thought which does attempt to disseminate information locally in a multitude of ways. Naturally, books like Milliken et al. [28], Rivera and Obón [e.g. 22 and other works by the same group] and Vandebroek et al. [8] are the most widely accessible works, but multiple examples using diverse media are now available.

Recently, the lack of projects which address the general public especially as it relates to the dissemination of scientific knowledge gathered about local and traditional medicinal plants (i.e. in the context of ethnopharmacology) has been criticized [29]. This is a continuing debate which has raged over many years, but we argue that in recent years considerable progress has been made and that some examples of how to achieve this are now available. The goal is to disseminate knowledge as widely as possible through a range of publications in rural or urban zones, to people with or without formal education, to children or the elderly. This article is a very personal account of how to achieve this and the authors' experiences, but there is an urgent need to assess the impact of such activities on a broader level, and, also to reassess the impact researchers have on the communities [cf. 30]. Of course, it is not possible to assess the impact quantitatively, but our clear impression in all field sites has been that the simple fact that such TKS are the focus of scientific investigation and are an essential element of giving renewed sociocultural value to such knowledge and that activities like the ones described here are of great interest to the communities we worked in.

References

1 Studley J: Dominant knowledge systems and local knowledge, 1998: Mtn-Forum Online Library Document, http://www.mtnforum.org/resources/library/stud98a2.htm (accessed August 2005).
2 Warren DM: Using Indigenous Knowledge in Agricultural Development. World Bank Discussion Papers, No 127. Washington, The World Bank, 1991.
3 Freeman MM: Appeal to tradition; different perspectives on Arctic wildlife management; in Brosted J (ed): Native Power. The Quest for Autonomy and Nationhood of Indigenous Peoples. Bergen, Universitetsforlaget, 1985.
4 Rivera D, Obón C, Inocencio C, Heinrich M, Verde A, Fajardo J, Llorach R: The ethnobotanical study of local Mediterranean food plants as medicinal resources in Southern Spain. J Physiol Pharmacol 2005;56(suppl 1):97–114.
5 Gadgil M, Berkes F, Folke C: Indigenous knowledge for biodiversity conservation. Ambio 1993;22:151–156.
6 Heinrich M, Rimpler H, Falk U, Antonio B: Plantas Medicinales Mixes: Remedios Caseros de San Juan Guichicovi, Oaxaca, México 1990.
7 Verde A, Fajardo J: Las plantas en la cultura popular de Castilla-La Mancha. Junta de Comunidades de Castilla-La Mancha. Toledo, Consejería de Educación, 2003, pp 1–102.
8 Vandebroek I, Evert T, AMETRAC: Plantas medicinales para la atención primaria de la salud. El conocimiento de ocho médicos tradicionales de Apillapampa (Bolivia). Cochabamba (Bolivia), Indústrias Gráficas Serrano, 2003.
9 The Local Food-Nutraceutical Consortium: Understanding local Mediterranean diets: a multidisciplinary pharmacological and ethnobotanical approach. Pharmacol Res 2005;52:353–366.
10 Rivera D, Obón C, Cano F, Robledo A: Introducción al mundo de las plantas medicinales en Murcia. Murcia, Ayuntamiento de Murcia, 1994.
11 Rivera D, Obón C: Las plantas, las esencias y los perfumes. Murcia, Ayuntamiento de Murcia, 1995.
12 Fajardo J, Verde A, Rivera D, Obón C: Las plantas en la Cultura Popular de la Provincia de Albacete. Albacete, Instituto de Estudios Albacetenses 'Don Juan Manuel', 2000.
13 Sánchez MD, García J, Gómez A, Zon S: Plantas útiles de la comarca de la Manchuela. Villamalea, Colectivo de Escuelas Rurales, 1994.
14 Verde A, Fajardo J: La Etnobotánica como recurso didáctico. El huerto escolar. CREA. Rev Cent Recurs Educ Ambient 2002;7:22–24.
15 Verde A, Fajardo J: La Etnobotánica en el currículo de Secundaria. Educar en el 2000 2003;7:52–55.
16 Verde A, Fajardo J, Rivera D, Obón C, Inocencio C, Heinrich M: Materiales Curriculares 'La alimentación en Castilla-La Mancha: de la escasez al desperdicio. (El valor de los alimentos locales y su utilización sostenible.)' Murcia, Azarbe, 2004.
17 Rivera D, Obón C: Guía de teoría y prácticas de Etnobotánica. Murcia, Diego Marín, 1998.
18 Chaparro P, Verde A, Fajardo J: Proyecto de Educación Ambiental para la Infancia 'El agua'. ASAJA, Fundación Biodiversidad y Consejería de Educación de la Junta de Comunidades de Castilla-La Mancha, Toledo (Multimedia) 2002.
19 Chaparro P, Verde A, Fajardo J: Proyecto de Educación Ambiental para la Infancia 'Flora y Fauna'. ASAJA, Fundación Biodiversidad y Consejería de Educación de la Junta de Comunidades de Castilla-La Mancha, Toledo (Multimedia) 2002.
20 Chaparro P, Verde A, Fajardo J: Proyecto de Educación Ambiental para la Infancia 'Agricultura sostenible'. ASAJA, Fundación Biodiversidad y Consejería de Educación de la Junta de Comunidades de Castilla-La Mancha, Toledo (Multimedia) 2002.

21 Verde A: El jardín escolar como recurso didáctico para atender a la diversidad. Educar en Castilla-La Mancha 2004;25:20–21.
22 Rivera D, Obón C: La Guía de Incafo de Las Plantas Útiles y Venenosas de La Península Ibérica y Baleares (Excluidas Medicinales). Madrid, Incafo, 1991.
23 Obón C, Rivera D: Las Plantas Medicinales de Nuestra Región. Murcia, Agencia Regional de medio Ambiente-Editora Regional, 1991.
24 Inocencio C, Rivera D, Obón C, Verde A, Heinrich M: Local food as a sustainable development issue in SE Spain. S Afr J Bot 2003;69:251.
25 Rivera D, Verde A, Fajardo J, Inocencio C, Obón C, Heinrich M (eds): Guía Etnobotánica de los Alimentos Locales Recolectados en la Provincia de Albacete. Albacete, Instituto de Estudios Albacetenses 'Don Juan Manuel', 2006.
26 Nebel S: Ta chòrta. Piante commestibili tradizionali di Galliciano. London, University of London, School of Pharmacy, 2005.
27 Cerchiaia L, Jannelli L, Longo F: Die Griechen in Süditalien: Auf der Spurensuche zwischen Neapel und Syrakus. Stuttgart, Konrad Theiss Verlag, 2004.
28 Milliken W, Bridgewater S: Flora Celtica. Plants and People in Scotland. Edinburgh, Birlinn Ltd, 2004.
29 Jaeger A: Is traditional medicine better off 25 years later? J Ethnopharmacol 2005;100:3–4.
30 Edwards S, Nebel S, Heinrich M: Methodological and epistemological considerations on the use of questionnaires in ethnopharmacology. J Ethnopharmacol 2005;100:30–36.

Michael Heinrich
Centre for Pharmacognosy and Phytotherapy
The School of Pharmacy, University of London
29–39 Brunswick Square, London WC1N 1AX (UK)
Tel. +44 20 7753 5844, Fax +44 20 7753 5909, E-Mail phyto@pharmacy.ac.uk

Heinrich M, Müller WE, Galli C (eds): Local Mediterranean Food Plants and Nutraceuticals. Forum Nutr. Basel, Karger, 2006, vol 59, pp 86–115

Plant Foods and Brain Aging: A Critical Appraisal

S. Schaffer, G.P. Eckert, S. Schmitt-Schillig, W.E. Müller

Institute of Pharmacology (ZAFES), Biocenter Niederursel, University of Frankfurt, Frankfurt, Germany

Abstract

In the 21st century, human aging will be one of the biggest challenges for most societies throughout the world. The decline in human fitness is a typical hallmark of the aging process. Aside from the cardiovascular system, the brain most often suffers significantly from the life-long impact of stressors, such as reactive oxygen and nitrogen species. Oxytosis, i.e. oxidative stress-induced cell death, has been identified to play a major role in the development and onset of chronic diseases. Foods, especially of plant origin, are rich in antioxidants and numerous in vivo data suggest that a diet rich in fruits and vegetables supports the maintenance of animal and human health. These beneficial effects also extend to the central nervous system, which, due to the presence of the blood-brain barrier, tightly controls the influx of metabolites and nutrients. In earlier studies the impact of antioxidant vitamins, such as α-tocopherol and ascorbic acid, on brain health has been of interest. Recently, the focus moved to assessing the potential of unsaturated fatty acids and secondary plant metabolites, particularly of polyphenols, to act as neuroprotectants. Considerable experimental evidence suggests that polyphenols and other plant-derived bioactivities affect animal and human brain function not only by directly lowering oxidative stress load but also by modulating various signal transduction pathways.

The phenomenon of aging is distinct from most other health alterations as it affects all members of a species. The aging process is irreversible and increases the vulnerability to certain diseases. A multitude of aging theories have been proposed based on either evolutionary, molecular, cellular and system-based considerations, and although the ultimate causes of aging remain unknown, recent evidence suggests that aging is presumably triggered by an interplay of intrinsic (genetic), extrinsic (environmental) and stochastic (random damage) parameters [1].

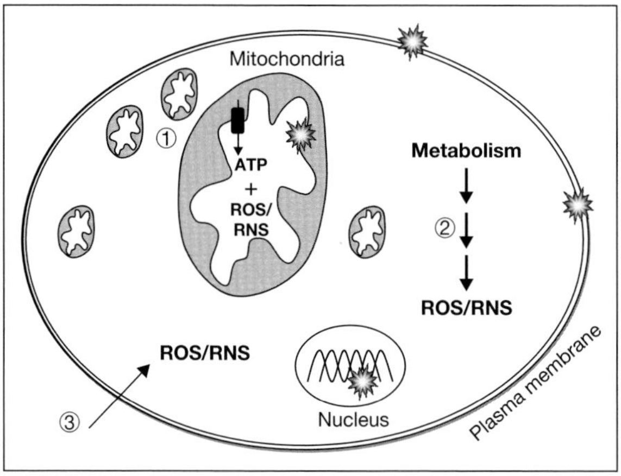

Fig. 1. An imbalance in the formation of reactive oxygen and nitrogen species (ROS and RNS) promotes the induction of oxidative stress. Mitochondria ① are both the major source of oxidants and at the same time key target or their deleterious effects. ROS and RNS are also generated during normal metabolic processes ② in the cytosol and other cell organelles. Furthermore, free radicals of exogenous sources ③, such as HOCl-producing immune cells, can damage cellular biomolecules, most notably fatty acids, proteins and DNA.

This chapter reviews possible pharmacological effects of nutrition-based interventions on one of the most ambitious goals in aging research, i.e. to facilitate healthy aging, particularly of the brain. Whereas the in vitro bioactivity of nutrients (e.g. unsaturated fatty acids, vitamins) and secondary plant metabolites (e.g. polyphenols, carotenoids) associated with healthy aging has been extensively reviewed in the past [2, 3], deficient knowledge still exists regarding the actual potential of these substances to alleviate aging-linked adverse health effects in animal and human brain.

Oxidative Stress Theory of Aging

The healthy human body tightly controls a cellular equilibrium between oxidizing and reducing equivalents. Oxidants consist of reactive oxygen and nitrogen species (ROS and RNS) either of free radical (e.g. $^{\bullet}OH$, $^{\bullet}OOH$, $O_2^{\bullet-}$) or non-radical (e.g. $ONOO^-$, H_2O_2, ^-OCl) nature. The occurrence of these reactive molecules is an attribute of normal aerobic life and some act as important messengers, such as NO, a potent vasorelaxant [4, 5]. An imbalance in the formation and removal of ROS and RNS, however, promotes the induction of oxidative stress (fig. 1), an event frequently implicated with the aging process itself as well as aging-related diseases. Biomarkers of ROS and RNS attack, such as lipid peroxides, protein carbonyls and damaged DNA molecules, are hallmark of cardiovascular disease (CVD), cancer, inflammation, dementia and various other chronic diseases, thus supporting the oxidative stress theory of aging and its associated morbidities [6].

Among all cellular organelles, mitochondria play the most prominent role in aging. Mitochondria are both a major source of oxidants and at the same time key targets for their deleterious effects [7]. Superoxide ($O_2^{\bullet-}$), for example, is formed when oxygen directly picks up electrons from complex III and complex I of the respiratory chain [8]. Thus, a mitochondrium produces about 10^7 highly reactive metabolites per day [9]. The loss of mitochondrial function, due to ROS- and RNS-mediated mtDNA, protein and lipid damage, has been repeatedly shown not only in apparently healthy aged animals and humans [10–12] but also in the onset of common diseases, especially CVD and neurodegeneration [8, 13]. Mitochondrial dysfunction is further characterized by a loss in mitochondrial membrane potential, an event subsequently leading to a reduction in cellular energy supply, and a drop in reduced glutathione (GSH), the most important endogenous house-keeping non-protein thiol antioxidant. Changes in the mitochondrial membrane potential in turn cause the augmented release of ROS and RNS, thus kicking off a vicious cycle of oxidative stress and tissue damage. It is now generally accepted that mitochondria not only take part in the genesis of cellular damage but also play a pivotal role in the complex pathway of apoptosis. Mitochondria embody both apoptotic cascade inhibitors (e.g. Bcl-xL, Bcl-2) and activators (e.g. Bax, cytochrome c). Recent data from animal studies suggest an age effect on mitochondrial gene expression linked to apoptosis in mice brain. Numata et al. [14], for example, found in brain of 3- and 12-month-old SAMP10 mice elevated amounts of cytochrome c whereas the protein mass of Bcl-2 was reduced compared to control animals. Similarly, Manczak et al. [15] observed increased cytochrome c levels and DNA damage in 12- and 18-month-old C57BL6 mice compared to 2-month-old animals.

Although all members of a species age, gender differences have been proposed. Premenopausal women, for example, are at decreased risk for CVD than males of similar age. High estrogen levels have been suggested to account for this observation, as estrogen participates in the modulation of mtDNA transcription, energy production and mitochondrial antioxidant status. The beneficial effect of estrogen, however, is lost after menopause resulting in a convergence of the risk for men and women to develop aging-related diseases [16].

Furthermore, increasing evidence indicates that not only gender differences but also early events in utero (termed 'fetal programming') have profound effects on aging and long-term health [17, 18]. Maternal hypercholesterolemia has particularly been associated with accelerated progression of arteriosclerosis in their offspring [19]. Recent data clearly suggest that core pathological mechanisms, especially the inflammatory components, promoting arteriosclerosis are also involved in the course of brain aging and severe neurodegenerative events, such as Alzheimer's disease (AD). This raises the question of whether certain environmental exposures (e.g. maternal nutrition) during the time in utero pose other possible risk factors for neurodegenerative events in normal brain aging and beyond [20].

Brain Aging

Human brain aging is typically accompanied by various histological modifications, such as shrinkage of the brain's gray and white matter volume, as well as biochemical and molecular alterations [21]. A recent DNA microarray study of human cortex and cerebellum revealed age-dependent changes in gene expression for both brain tissues. However, not only did many more genes undergo expression changes in the cortex than the cerebellum, the gene expression profiles also differed significantly from those obtained from aged chimpanzee brain, suggesting distinct species differences [22]. Furthermore, already early in the aging process, Ca^{2+} homeostasis in the brain begins to be dysregulated. Changes in cellular Ca^{2+} pools in turn affect the activity of mitochondria and subsequently ROS and RNS formation [23, 24]. The brain is considered to be especially prone to oxidative stress due to its high oxygen consumption. Whereas adult human brain accounts for less than 2% of the body weight, it processes about 20% of basal oxygen consumption, mainly to produce energy (ATP) via mitochondria. The high number of mitochondria per neuronal cell as well as the high amount of polyunsaturated fatty acids (PUFAs), which are prone to undergo oxidative modifications, furthermore contribute to the increased vulnerability of the brain to oxidative stress [25, 26].

Although the brain contains about 10^{11}–10^{12} neurons and approximately twice as many glia cells, neuronal cell death is particularly troublesome as de novo formation of neurons only proceeds at marginal levels [9, 27]. Therefore, a wide spectrum of memory and cognitive loss occurs from normal brain aging to mild impairment, to severe maladies such as Parkinson's disease and AD [28]. Aside from alterations in cognition, neurological aging often also impairs motor skills, thus hampering an individual's ability for a self-determined life (fig. 2).

In order to prevent oxidative damage, the cell has evolved a number of synergistic defense mechanisms to deal with oxidative stress and oxidative damage. The antioxidant enzymes – superoxide dismutase (SOD), catalase (CAT), glutathione peroxidase (GPx) and glutathione reductase (GR) – form the first line of defense. The enzymes act in concert to remove ROS and RNS. Whereas superoxide dismutases (SOD-1, SOD-2) convert $O_2^{\bullet-}$ to H_2O_2, GPx and CAT

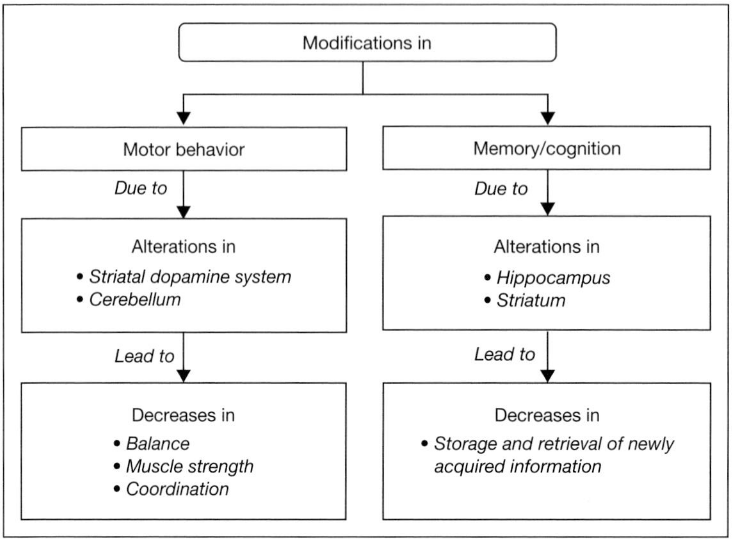

Fig. 2. Causes and consequences of aging on cognitive performance and motor performance.

are responsible for the removal of H_2O_2. Although the detoxifying activity of CAT increases in the presence of high H_2O_2 concentrations, GPx seems to be more important for the protection of tissues, especially in the brain where CAT activity is comparatively low [29, 30].

Further antioxidant support comes from endogenous (e.g. GSH, uric acid) and exogenous (e.g. vitamin C, secondary plant metabolites) low molecular weight ROS and RNS scavengers. These molecules are unevenly localized in vivo. Whereas hydrophobic antioxidants are present in the lipophilic phase of biological membranes, hydrophilic ones accumulate in the cytosol and extracellular fluids [6, 31, 32]. Albeit not clear whether oxidative stress is cause or consequence of neurodegeneration, the insufficient removal of ROS and RNS significantly triggers a set of deleterious downstream events: (i) mitochondrial dysfunction; (ii) glial cell activation; (iii) proteasomal malfunction, and subsequently (iv) apoptosis (programmed cell death).

Data from various studies assessing ROS- and RNS-dependent biomarkers (e.g. nitrotyrosine, NT; malondialdehyde, MDA) indicate that there is an increase in oxidative stress vulnerability as a function of age. In several brain regions of adult (4- to 6-month-old) rats, for example, no NT-immunoreactive cells were found whereas NT immunoreactivity was significantly elevated in the brain of aged (24- to 29-month-old) rats [33]. Furthermore, Leutner et al.

[34] detected in young, adult, and aged mice steadily increasing amounts of MDA, suggesting an age effect on lipid peroxidation. Decreasing MDA levels with age, however, have recently been reported by our group using another non-transgenic mouse strain, again pointing to species differences in aging-related biomarkers [35]. Changes in the efficiency of the endogenous antioxidant network, i.e. modifications in the activity of SOD, GPx, and GR as well as the amount of direct scavengers, such as GSH, have been detected in aging subjects, too [34, 36, 37]. Comparable modifications in the antioxidant network and the load of oxidative stress-related biomarkers have been reported for human and other primate brain samples, particularly when severe neurodegeneration occurs [38–41].

As previously stated, cognitive function shows many alternative outcomes during aging. It is now widely accepted that food and nutrition, in addition to an individual's genetic setting, significantly affect the status of brain health.

Impact of Nutrition on Aging-Related Adverse Health Effects

Increasing evidence from epidemiological studies show that a regularly high dietary intake of fruits and vegetables is associated with reduced burden of disease [42, 43]. In the Mediterranean basin, particularly in rural areas, wide parts of populations consume a daily diet rich in foods of plant origin [44–46]. The health-beneficial effects of such a dietary habit were first recognized in the 1950s by the epidemiologist Ancel Keys (1904–2004), who later coined the concept of the Mediterranean diet as a nutrition-based approach for maintaining human well-being. Keys was the first to link the vast differences in CVD mortality between American and certain European populations to major dietary differences. Consequently, Keys and co-workers initiated the so-called 7-Countries-Study, a longitudinal assessment of anthropometrical, nutritional and disease parameters in 16 male cohorts originating from 5 European countries, the USA and Japan [47]. The primary outcome of this research, i.e. the link between low dietary intake of saturated fatty acids as well as high ingestion of flavonoids and reduced mortality caused by cardiovascular events [48], provoked subsequent intervention studies in the past decades. Exchanging the usual Finnish diet for a Mediterranean type of diet, for example, led to a 20% decrease in LDL cholesterol in middle-aged men and women [49]. Additionally, the Lyon Diet Heart Study revealed that consuming a modified diet of Crete significantly contributed to secondary prevention of CVD in patients with previous myocardial infarction [50]. In a recent intervention trial, Ambring et al. [51] found positive effects of a Mediterranean-inspired diet on blood lipids (i.e. triglycerides, cholesterol and LDL) of healthy subjects whereas vascular function and oxidative stress parameters remained unchanged. The latter observations might be due to the rather young age of subjects (average 43 years) as well as their overall low-risk profile for CVD. Further proof for a health-beneficial impact of a fruit- and vegetable-rich diet derives from the EPIC trial, a multicenter, prospective cohort study in nine European countries, including almost 75,000 men and women aged 60 or more [52]. The authors showed for the first time a significant reduction in overall death rate resulting from a diet relying on plant foods and unsaturated fatty acids.

Fatty Acids in Brain Health and Aging

Epidemiological and clinical secondary prevention trials suggest that the omega–3 polyunsaturated fatty acids (n–3 PUFAs) in Mediterranean diets may play a significant role in the prevention of CVD [53, 54]. In contrast, information on the effects of Mediterranean diet and especially of PUFAs on neuronal function is still incomplete. However, a recent review emphasized that specific macronutrients of the Mediterranean diet may act synergistically with other protective factors, reaching to new therapeutic interventions for cognitive decline [55, 56]. Monounsaturated fatty acids (MUFAs) and PUFAs represent macronutrients of the Mediterranean diet that have been studied for central nervous system (CNS) function and neuropsychiatric diseases [57].

PUFAs provided the first coherent experimental demonstration of the effect of diet (i.e. nutrients) on the structure and function of the brain. It has been shown that PUFAs could prevent some aspects of CVD. PUFAs also seem to be involved in some neuropsychiatric disorders, particularly depression, as well as in brain aging, dementia and AD. The brain, after adipose tissue, is the organ richest in lipids [58]. Its dry weight consists of approximately 50% of lipids, 35% of which are PUFAs. Arachidonic acid (AA) and docosahexaenoic acid (DHA) represent the dominant fatty acid species [59]. PUFAs cannot be synthesized de novo but can be formed from linoleic acid (LA) and α-linolenic acid (ALA) (fig. 3). Primrose, sunflower, corn or safflower oil are rich in LA, while hemp, flax or fish oil, as well as nuts and green leafy vegetables, such as purslane, are main sources of ALA [59–61]. The estimated daily intake of n–3 PUFAs in Western countries varies largely, but is often below the recommended dietary allowance. Mean DHA intake, for example, mainly depends on fish consumption, which differs greatly between countries [57].

By utilizing the two essential unsaturated fatty acids LA and ALA, the microsomal enzyme systems of mammalian cells desaturate and lengthen the molecules progressively to form highly unsaturated, long-chained fatty acids (fig. 3) [59]. However, in contrast to other cell types, neurons appear to be

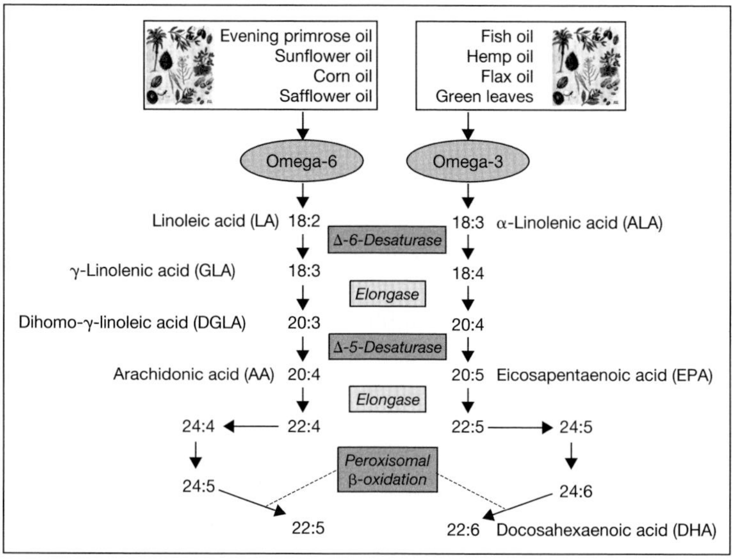

Fig. 3. Metabolism of omega–3 and –6 polyunsaturated fatty acids [59, 148, 149].

unable to carry out fatty acid desaturation and thus depend on preformed long-chain PUFA supply [62, 63]. Intracellular concentrations of non-esterified fatty acids and fatty acid acyl-CoA are below 10 μ*M*, a major fraction of which is bound to specific proteins, i.e. fatty acid-binding proteins. Furthermore, fatty-acid-CoAs are substrates for neutral lipids, such as cholesterol esters and triglycerides, and polar lipids, such as phospholipids [64]. The various metabolic products of dietary fatty acids are essential for physiological functions in the body, especially the central and peripheral nervous system [59]. Moreover, PUFAs have been shown to participate in the formation of the blood-brain barrier by inducing the assembly of tight junctions that create a rate-limiting barrier to the diffusion of solutes between vertebrate epithelial cells [65]. Like cholesterol, PUFAs modulate many of the signal transduction mechanisms by changing the physicochemical properties of neuronal cell membranes: PUFAs decrease the rigidity and thereby increase the fluidity of biological membranes [59, 66]. Normal physiological functions of neurons include the transmission of axonal information, regulation of membrane-bound enzymes and control of ion channels and various receptors all of which are highly dependent on membrane fluidity [67]. Major biochemical changes in the brain affect neuronal membranes and influence essential brain functions, such as long-term potentiation, learning and memory, sleep, pain threshold and thermoregulation [56].

Various neurotransmitters, such as serotonin, acetylcholine or the catecholamines, interact with members of a heptahelical transmembrane receptor family that are G-protein-coupled [59]. Transport proteins can also be modulated by PUFAs. DHA, for example, inhibits Ca^{2+}-ATPase in neuronal membranes [59]. A decrease in the ALA series in membranes results in a 40% reduction in the Na^+, K^+-ATPase activity of nerve terminals and a 20% reduction in 5′-nucleotidase activity. However, not all enzymatic activities are affected by altered membrane fluidity [68]. Small changes in the composition of dietary fat intake can modify the composition of brain membranes during development [69] and during the aging process [70]. Table 1 summarizes recently published PUFA effects on brain function in animal models. Aside from their impact on physicochemical properties of biological membranes, PUFAs also influence brain gene expression. Two studies by Kitajka et al. [71, 72] demonstrated membrane-independent effects of PUFAs on brain gene expression in aged rats. Using a mouse model, Puskas et al. [73] showed that a diet enriched in cholesterol and DHA induces gene expression changes in encoding of transcription factors and regulators, fatty acid-binding proteins as well as inflammatory proteins.

Whereas promising data exist with respect to constitutional effects of PUFAs on behavior, learning and cognition in pups, toddlers and children [74–78], their impact on human brain function during aging is less clear. Still, epidemiological and small-scale intervention studies on the association between diet and cognitive capacity suggest a possible role of PUFAs in maintaining adequate cognitive function and possibly in preventing or delaying the onset of cognitive decline [79, 80].

Classical Health Modulators: Vitamin E and Vitamin C

About 30 years after the discovery of the importance of vitamin E (tocopherols and tocotrienols) on reproduction in 1922, research shifted the focus on the free radical scavenging and chain-breaking activity of this lipophilic nutrient [81]. Recently, the role of vitamin E as a signal transduction molecule, particularly in the brain, has been reported [82, 83].

Based on several epidemiological studies, the health benefits of vitamin E in humans have been proposed [84–86]. Consequently, animal studies have been carried out to elucidate the underlying mechanism responsible for presumed effects on human well-being (table 1).

In the past decades, several groups investigated the impact of vitamin E on human health using intervention trials. In contrast to the animal studies mentioned in table 1, vitamin E supplementation in humans showed no or marginal positive effects not only on CVD [87, 88] but also on cognitive performance and

Table 1. In vivo effects of single food constituents on antioxidant and non-antioxidant parameters in animal brain

Reference	Intervention	Species	Dosage	Effect on		
				oxidative stress biomarker	other health parameters	motor & cognitive behavior[1]
Polyphenols						
[150]	Apigenin-7 -glucoside & Quercetin	Mouse	Apigenin: 5–20 mg/kg bw, Quercetin: 25–100 mg/mg bw (i.p.; daily; 7 days)	N/A	N/A	Dose-dependent reversal of age- and LPS-induced retention deficits (passive avoidance & elevated plus maze test) found for both flavonoids
[151]	Curcumin	Rat	300 mg/kg bw (i.p.; 30 min after MCAO)	Brain: lipid peroxidation ↓, GPx ↑, proxynitrite ↓	Brain: infarct volume ↓, cerebral edema ↓	N/A
[152]	Epigallocatechin gallate (EGCG)	Gerbil	2 × 50 mg/kg bw (i.p.; 30 min before & immediately after ischemia)	Postischemic brain: MDA ↓	Postischemic brain: edema & infarct volume ↓	N/A
[153]	Epigallocatechin gallate (EGCG)	Rat	50 mg/kg bw (i.p.; after ischemia)	Postischemic brain: MDA ↓, GSSG to GSH ratio: ↓	Postischemic brain: infarct volume ↓	Neurological deficit score: ⊕
[154]	Naringenin	Mouse (adult)	4.5 mg/kg diet (oral; daily; 3 weeks prior to scopolamine-induced amnesia)	N/A	N/A	Latency time (passive avoidance test) ⊕; alternation behavior (y maze) ⊕

Table 1. (continued)

Reference	Intervention	Species	Dosage	Effect on		
				oxidative stress biomarker	other health parameters	motor & cognitive behavior[1]
[155]	Quercetin	Rat	200 mg/kg bw (oral; 2 h prior to LPS treatment)	Brain: MDA ↓, nitrite/nitrate ↓, phospholipids ↑ Plasma: sulfhydryl groups ↑	N/A	N/A
[156]	Resveratrol	Rat	30 mg/kg bw (oral; daily; 5 days)	N/A	Brain: KA-induced neuronal (hippocampus) damage ↓; astrocyte & microglia activation ↓	N/A
[157]	Resveratrol	Rat	10–20 mg/kg bw (i.p.; daily; 21 days; days 1 and 3: STZ injection)	Brain: MDA ↓, GSH ↑	N/A	Retention latency (passive avoidance test): ⊕; Transfer latency (elevated plus maze test): ϟ
[158]	Rutin	Rat	40 mg/kg bw (oral; daily; 2 weeks + 3 weeks of exposure)	TEAC after smoke exposure: plasma ↑, brain ↓	N/A	N/A
[159]	Wogonin	Rat	20 mg/kg bw (i.p.; 30 min before & 4 h after MCAO)	Brain: infarct area & volume ↓	N/A	Neurological score (degree and duration of forelimb flexion + forepaw outstretching) ⊕

Vitamins						
[160]	Ascorbic acid	Mouse (adult)	60 mg/kg bw (i.p; daily; 3 days)	N/A	N/A	Transfer latency (elevated plus maze test): ↯; Step-down latency (passive-avoidance apparatus): ⊕
[161]	Ascorbic acid	Mouse	266 mg/kg bw (i.p.; daily; 7 days)	Brain: TBARS ↑	N/A	N/A
[162]	Ascorbic acid	Rat (adult)	15 mg/kg bw (30 min prior to or post 6 h immobilization)	Brain: SOD ↑, CAT ↑, GST ↑, GSH ↑, MDA ↓	N/A	N/A
[163]	Ascorbic acid	Monkey (adult)	500 mg/kg bw (5–10 min prior to 4 h occlusion of major cerebral blood vessel and subsequent reperfusion)	N/A	Brain: infarct size ↓	N/A
[162]	Vitamin E (α-tocopheryl acetate)	Rat (adult)	15 mg/kg bw (30 min prior to and post 6 h immobilization)	Brain: SOD ↑, CAT ↑, GST ↑, GSH ↑, MDA ↓	N/A	N/A
[164]	Vitamin E	Rat (aged)	No conc. specified (oral; daily; from 4th month onwards)	Brain: GPx ↑, SOD ↑, CAT ↓, MDA ↑, lipofuscin ↓	N/A	N/A
[165]	Vitamin E	Rat (adult)	500 IU/kg bw (oral; daily; 2 months)	Brain: protein carbonyl ↓	Brain: BDNF expression ↑; CREB expression ↑	Escape latency (Morris water maze) ⊕; Memory retention (spatial probe test): ⊕

Table 1. (continued)

Reference	Intervention	Species	Dosage	Effect on		
				oxidative stress biomarker	other health parameters	motor & cognitive behavior[1]
Unsaturated fatty acids						
[166]	n–3 fatty acids (Marincarp capsule©)	Rat (adult)	400 mg/kg bw (oral; daily; 14 days prior to formaldehyde exposure)	Brain: MDA ↓, GPx ↑, SOD ↑, number of BAX-stained cells ↓	N/A	N/A
[167]	DHA (+ palm oil)	Mice	DHA: 2% of diet (oral; daily; from 3rd week of age onwards for 7 months)	N/A	Brain: DHA ↑, AA ↓	Latency (maze test): ⊕
[168]	n–3-rich fish oil	Rat	Fish oil: 14% of diet (oral; daily; from 4th week of age onwards for 6 weeks prior to ischemia)	Brain (after ischemia): TBARS ↑, CAT ↑, SOD ↑	Brain: infarct volume ↓	N/A

AA = Arachidonic acid; bw = body weight; CAT = catalase; DHA = docosahexaenoic acid; GPx = glutathione peroxidase; GSH = glutathione; GSSG = gluthadione disulfide; GST = gluthadione-S-transferase; KA = kainic acid; LPS = lipopolysaccharide; MCAO = middle cerebral artery occlusion; MDA = malondialdehyde; N/A = not applicable; SOD = superoxide dismutase; STZ = streptozotocin; TBARS = thiobarbituric acid reactive substances; TEAC = trolox equivalent antioxidant capacity.

[1]Improvement = ⊕; impairment = ↯.

memory. Whereas modest benefits from vitamin E supplementation have been reported by Sano et al. [89], Petersen et al. [90] found no benefit of 3-year supplementation with 2,000 IU vitamin E daily in patients with mild cognitive impairment. The outcome of a recent mice feeding trial, however, supports the potential of vitamin E as a neuroprotectant and memory enhancer during aging. Aged male mice receiving vitamin E from 28 weeks of age showed not only a significantly increased lifespan but also better performance in neuromuscular function (i.e. tightrope) and exploratory activity (i.e. T-maze) tests. Concomitantly, vitamin E supplementation partially ameliorated brain mitochondrial dysfunction as well as loss in antioxidant enzyme activity [91]. This and other studies also indicate that the duration of vitamin E supplementation affects study outcome: short-term supply of elevated vitamin E doses failed to improve cognitive or psychomotor performance of aged mice [92]. Furthermore, the apparent lack of impact of vitamin E intake on health has been attributed by some researches to the fact that most studies to date have been assessing the effects of α-tocopherol. Although α-tocopherol possesses the highest biological activity and, due to the presence of a specific binding protein, is the predominant form of all vitamin E derivatives in plasma and certain body tissues, γ-tocopherol represents the most abundant tocopherol in human diets and exerts unique biological effects, such as inhibition of cyclooxygenase [93, 94]. Moreover, some studies revealed an inverse link between γ-tocopherol and the incidence of chronic diseases [95]. Data from Vatassery et al. [96] indicate that γ-tocopherol can enter the brain. Also, γ-tocopherol and other tocopherol forms rather than α-tocopherol alone may be implicated in vitamin E's protective association with AD [97]. Besides γ-tocopherol, the tocotrienols have also been suggested as potential neuroprotectants [98, 99].

The impact of vitamin C on health parameters has been extensively studied, as well, due to its excellent in vitro antioxidant activity and ubiquitous natural occurrence in plants. Vitamin C acts as a strong hydrophilic antioxidant because of its reducing properties and affects cell physiology by modulating signal transduction pathways as well as neurotransmitter synthesis [100]. As part of the antioxidant network, vitamin C scavenges a broad spectrum of reactive metabolites and regenerates other antioxidants such as vitamin E or β-carotene [101]. Based on data from animal and human studies, there appears to be a higher requirement for vitamin C during aging due to a decline in the capacity of cells to absorb this antioxidant [102, 103] and its increased consumption [104]. Vitamin C, entering the brain in its oxidized form, i.e. dehydroascorbic acid, via facilitated transport [105], reaches neuronal concentrations of up to 10 m*M* [106]. Neuroprotection by vitamin C has been demonstrated in several recent animal studies (table 1) [101, 107]. As with vitamin E, while there are only few investigations that have shown a positive link between vitamin C intake and cognitive performance [108, 109], there are other published data that contradict such beneficial effects [110, 111]. Hence, the apparent inconsistencies regarding the effect of isolated vitamin C and E supplementation on brain aging are far from being resolved. Encouraging news come from recent studies investigating the combined effect of vitamin C and E on cognitive decline. Data from the Nurses' Health Study, the Honolulu-Asia Aging Study as well as the prospective, 5-year Canadian Study of Health and Aging indicate better cognitive performance in women and men and a reduced risk to experience cognitive decline, respectively, in humans with a combined use of vitamins C and E [108, 110, 112].

Putative Novel Neuroprotectants: Polyphenols

The partly disappointing results obtained in intervention studies using antioxidant vitamins as well as the so-called 'French paradox', i.e. the apparent compatibility of a high-fat diet with a low incidence of CVD [113], directed the interest to a further class of potentially health-beneficial plant constituents: the polyphenols (PPs). Widely distributed in the leaves, seeds, bark, and flowers of plants, several thousand PPs (e.g. flavonoids, coumarins, stilbene) have been identified to date [114]. Over the past decades, PPs have been shown in vitro to act as efficient antioxidants, to affect the activity of various enzymes, to possess hormonal-like activity, to modulate gene expression and to interact with several cellular receptors, thus indicating potential as anticarcinogenic, anti-inflammatory, and antineurodegenerative substances [115]. Moreover, PPs are capable of increasing vitamin E concentrations in vivo [116]. While PPs have been shown to reach low to medium micromolar concentrations in the circulatory system of humans and animals after oral intake [117–122], only a small number of studies have investigated the actual penetration of PPs into the brain of supplemented subjects (table 2). No complete picture of the in vivo bioactivity of PPs has been drawn, but more and more promising data emerge indicating that PPs might indeed not only influence the aging of the body periphery [123] but also of the brain. As summarized in table 1, a substantial number of animal studies showed that PPs have the potential to alleviate the levels of oxidative stress-related biomarkers, an effect that possibly contributes to the concomitantly observed improvement in cognitive performance. Interestingly, the latter effects have not only been detected in adult and hence rather healthy, but also in aged laboratory animals. In contrast, no data exist regarding the effect of PP supplementation on brain function in humans. Observational studies, however, suggest a link between dietary flavonoid intake and a reduced risk for the development of dementia [124].

Table 2. Presence of polyphenols in animal brain

Reference	Species	Dosage	Outcome
[169]	Rat	15 g blackberry extract/kg diet (oral; daily; 15 days)	Total anthocyanins: 0.25 ± 0.05 nmol/g tissue
[170]	Rat	Quercetin: 1% of diet (oral; daily; 11 weeks)	Quercetin and its metabolites: 0.68 nmol/g tissue
[170]	Pig	Quercetin: 500 mg/kg bw (oral; daily; 3 days)	Quercetin and its metabolites: 0.22 nmol/g tissue
[171]	Rat	2% blueberry (oral; daily; 8–10 weeks)	Several anthocyanins detected in brain tissue

In vivo Effect of Dietary Plant Foods on Brain Aging

Risk factors for cognitive decline, a common occurrence in old age, are multivariate, hence the potential of several possible prevention strategies, including the use of antioxidant-rich foods or food supplements, are currently being tested. A plant-based diet is thought to have played a significant role in human evolution [125] and the consumption of whole plant foods (fruits, vegetables, legumes and nuts) or their extracts (e.g. tea or grape seed extract) has been repeatedly associated with a decreased risk for aging-related diseases. The impact of dietary intervention with green leafy vegetables (i.e. spinach), fruits (i.e. blueberry and strawberry), spices (i.e. garlic), tea, or grape (including red wine) on biochemical markers and cognitive parameters in animals is summarized in table 3. Moreover, secondary plant metabolite-rich extracts prepared from non-nutritional plants have also been found to prevent aging-related increase in oxidative stress and to improve the cognitive performance in aged animals [126, 127]. Furthermore, in women aged 60 years and above, Lee et al. [128] observed that subjects of poor cognitive performance in the Mini-Mental State Examination had significantly lower intakes of vegetables, fruits and spices that those of the normal score group. Based on data from the Nurses' Health Study, more than 15,000 women were recruited by Kang et al. [129] for a prospective assessment of fruit and vegetable intake in relation to cognitive function. In contrast to fruit intake, total vegetable consumption was significantly associated with less cognitive decline. The strongest association was found with greater intake of green leafy vegetables (e.g. spinach and romaine lettuce) and cruciferous vegetables (e.g. broccoli and cauliflower).

Mediterranean diets, especially the one traditionally practiced in Crete, are characterized by a very high alimentary intake of greens [130]. Here, a

Table 3. In vivo antioxidant and non-antioxidant effects of fruits and vegetables and/or their derived beverages or extracts on animal brain

Reference	Intervention	Species	Dosage	Effect on		
				oxidative stress biomarker	other disease parameters	motor & cognitive behavior[1]
[172]	Berries (blueberry)	Rat (adult)	2% of diet (oral; daily; 4 weeks prior to ischemia)	Brain: cell death ↓	Brain: infarct volume ↓	Post-stroke locomotor activity: ⊕
[173]	Berries (blueberry)	Rat (aged)	Extract: 18.6 g/kg diet (oral; daily; 8 weeks)	Brain: ROS level ↓	N/A	Working memory (Morris water maze): ⊕; Psychomotor coordination (rod walking and rotarod): ⊕
[174]	Berries (blueberry)	Rat (aged)	Extract: 18.6 g/kg diet (oral; daily; 8 weeks)	Brain: GSH ↑	Brain: β-adrenergic receptor function ↑	N/A
[174]	Berries (strawberry)	Rat (aged)	Extract: 14.8 g/kg diet (oral; daily; 8 weeks)	Brain: GSH ↑	Brain: β-adrenergic receptor function ↑	N/A
[175]	Garlic	Mouse (SAMPs)	Extract: 2% of diet (oral; daily; 8–9 months)	N/A	Longevity (SAMP 8): ↑, brain atrophy: ↓ (SAMP 10)	Passive avoidance: ⊕ (SAMP 8); conditioned avoidance: ⊕ (SAMP 8); escape latency (water maze): ϟ(SAMP 8)
[176]	Grape (seed)	Gerbil (adult)	Extract: 5 g/kg diet (oral; daily; 2 months)	Brain: cell death ↓, microglia activation: ↓, DNA damage: ↓	N/A	N/A
[177]	Grape (seed)	Rat (aged)	Extract: 100 mg/kg bw (oral; daily; 30 days prior to ischemia)	Brain: ROS ↓, protein carbonyl: ↓, thiols: ↑	N/A	Memory (T maze): ⊕

[178]	Grape (seed)	Rat (aged)	Extract: 100 mg/kg bw (oral; daily; 30 days prior to ischemia)	Brain: MDA ↓, SOD ↑, CAT ↑, GPx ↑, GSH ↑, vitamin C ↑, vitamin E ↑	N/A	N/A
[172]	Spinach	Rat (adult)	2% of diet (oral; daily; 4 weeks prior to ischemia)	Brain: caspase-3 activity ↓, cell death ↓	Brain: infarct volume ↓	Post-stroke locomotor activity: ⊕
[174]	Spinach	Rat (aged)	Extract: 9.1 g/kg diet (oral; daily; 8 weeks)	Brain: GSH ø	Brain: β-adrenergic receptor function ↑	Learning (rod-running motor task): ⊕
[179]	Tea (green)	Mouse (SAMP 10)	Catechins: 0.02% in drinking water (ad libitum; 11 months)	DNA damage in rhinencephalon: ↓ (6 months of age), ø (12 months of age)	Cerebral atrophy: ↓ (12 months of age)	Learning: ⊕ (11 months of age); memory retention: ⊕ (12 months of age)
[180]	Tea (green)	Gerbil	Extract: 2% in drinking water (ad libitum; 3 weeks prior to ischemia)	Brain: hydrogen peroxide ↓, MDA+HNE ↓, DNA damage ↓, cell death ↓	Brain: infarct volume ↓	N/A
[181]	Tea (unspecified polyphenols)	Mouse	0.2% of diet (7 weeks prior to scopolamine-induced retention deficit)	N/A	Brain: AChE activity ↓	Latency time (passive avoidance test) ⊕; alternation behavior (Y maze): ⊕

AChE = Acetycholine esterase; CAT = catalase; GPx = glutathione peroxidase; GSH = glutathione; HNE = hydroxynonenal; MDA = malondialdehyde; N/A = not applicable; ROS = reactive oxygen species; SOD = superoxide dismutase.

[1]Improvement = ⊕; impairment = ↯; no change versus control = ø.

Table 4. Activity (selection) of promising extracts prepared from local Mediterranean plant foods in a free radical scavenging assay (DPPH), a hypochlorous acid-induced oxyhemoglobin bleaching assay (OxyHb), a xanthine oxidase assay (XO), a myeloperoxidase-catalyzed guaiacol oxidation assay (G-OH), an antiproliferation assays (BrdU) and an antidiabetic assay (PPARγ) [for more details, see 133]

Plant food	Activity, %					
	DPPH	OxyHb	XO	G-OH	BrdU	PPARγ
Berberis vulgaris	≥50	50–25	≥20	>75	<30	>75
Lythrum salicaria	30–39	>50	<10	>75	n.d.	>75
Reichardia picroides	<30	>50	≥20	>75	>50	>75
Scandix australis	<30	50–25	≥20	>75	30–50	>75
Satureja montana	<30	50–25	≥20	>75	n.d.	<30
Thymus piperella	30–39	>50	≥20	>75	30–50	<30
Vitis vinifera	≥50	50–25	≥20	>75	n.d.	<30

n.d. = Not determined.

considerable part of the ingested vegetables originates from semicultivated or wild sources. Interestingly, these plants often show a very high antioxidant activity, an effect that has been partly attributed to their particularly high polyphenol content [131, 132]. The EU-funded project 'Local Food-Nutraceuticals' aimed to further elucidate potentially health-beneficial effects of locally grown food plants of the Mediterranean basin [46]. A subset of a total of 127 plant extracts, i.e. *Berberis vulgaris*, *Lythrum salicaria*, *Reichardia picroides*, *Scandix australis*, *Satureja montana*, *Thymus piperella* and *Vitis vinifera*, showed promising activity profiles in diverse activities such as free radical scavenging, enzyme inhibition or prevention of angiogenesis (table 4). The identification of several plant extracts significantly preventing the occurrence of lipid peroxidation and ROS levels in mouse brain homogenate is of particular interest in face of potential neuroprotection [133, 134]. Table 5 summarizes in vitro and in vivo (mice feeding trial) effects of *Reichardia picroides* extract on brain aging-related parameters. As previously mentioned, epidemiological data suggest an inverse correlation of plant food intake and healthy brain aging, an effect that has been linked to both the consumption of fruits and vegetables in general as well as biological effects of certain plant constituents, such as flavonoids [124, 129]. However, although flavonoids react rapidly with ROS and RNS in test tube and cell culture models, it is questionable whether the levels of PPs obtained in human tissue are high enough to significantly enhance the antioxidant status of the body periphery let alone the brain [135]. Hence, the observed effects of fruits and vegetables on brain function remain unaccounted for. A series of recent reviews, however, indicate that PPs are more than

Table 5. Effects of an extract prepared from the local Mediterranean plant food *Reichardia picroides* on brain aging-related parameters assessed in vitro (mouse brain homogenate) and aged NMRI mice ([1]brain, [2]plasma, ex vivo) after a 3-month feeding trial with 100 mg extract/kg bw [S. Schmitt-Schillig, unpubl. observation]

In vitro		Ex vivo				
ROS level	lipid peroxidation	ROS level[1]	Mitochondrial membrane potential[1]	DNA damage[1]	TEAC[2]	hydro-peroxides[2]
↓	↓↓	↓	↑	↓	↑	↓

pure antioxidants. Mandel et al. [136], for example, developed a model to illustrate the neuroprotective activity of catechins based on their modulation of signal transduction pathways (e.g. PKC, PKA, MAPKs), the amount of protective/deleterious proteins (e.g. Bcl-2, BAX, APP) and mitochondrial activity. Similar properties have been described for resveratrol and flavonoid-rich extracts such as *Ginkgo biloba* [137, 138]. In addition, interaction of flavonoids with receptors regulated by neurotransmitters might explain some of the effects observed in vitro and in vivo [139, 140]. Other features of Mediterranean diets, such as the high intake of PUFAs, are also attracting more and more scientific interest. Wild purslane (*Portulaca oleracea*), an excellent source of ALA and antioxidants, is widely used not only in the Mediterranean basin but also the Asian tropics, Africa and various other countries. Extracts from purslane possess neuropharmacological effects, some of which have been suggested to be due to the presence of ALA [141, 142]. A recent review by Yehuda [56] provides an excellent overview on promising brain-related functions of PUFAs in animals and humans.

Finally, brain aging should not be looked at as an isolated case as mounting data indicate a strong interplay of the integrity of the body periphery with the CNS. CVD, for example, appears to contribute to cognitive decline [143, 144]. Consequently, lowering the burden of pathological events contributing to arteriosclerosis and other CVD-linked parameters could help maintain a functioning brain in old age. Several extracts derived from local Mediterranean plant foods have been found to prevent HOCl-induced, atherogenic modifications of LDL in vitro. Furthermore, the formation of 3-nitrotyrosine, a biomarker for MPO-mediated NO_2^- oxidation, has been reduced in the presence of plant extracts prepared from *Cynara cardunculus* and *Thymus pulegioides* [145]. Also, in elderly but otherwise healthy women, consumption of spinach, strawberries or alcohol-free red wine was shown to increase serum antioxidant capacity (SAC). Interestingly, vitamin C originating from the ingested plant sources as well as changes in urate status could only partly explain the improved SAC, suggesting that other plant food constituents may contribute to the observed SAC elevation [146]. A recent review of numerous human experimental studies highlights the positive effect of fruit and vegetable interventions on various disease-linked biological markers, such as immune cell activation, platelet aggregation, cholesterol levels and blood pressure [147]. Most probably, these health-beneficial effects in turn affect brain aging.

Conclusions

The advent of modern medicine and industrial-scale production of staple food have allowed more and more people to live for eight or more decades. Along with the rise in life expectancy, the number of people suffering from impaired brain function has dramatically increased. In addition to the genetic predisposition, exposure to environmental pollutants, and psychosocial influences, it is now generally accepted that an individual's food consumption pattern throughout the life cycle has a major impact on aging and the development of chronic diseases. Wild plant foods have contributed to the daily diet since the advent of humankind, and whereas numerous Western societies exchanged more and more plant foods with animal products, most Mediterranean populations, especially those in rural areas, maintained a largely plant-based diet. Convincing evidence exists in terms of reduced risk to develop CVD in conjunction with Mediterranean diets. Recent data from animal experiments as well as epidemiological studies indicate that interventions with plant foods also have a beneficial effect on brain damage and malfunction during aging [182]. The promising bioactivity profile of several groups of secondary plant metabolites, especially PPs (e.g. flavonoids, phenolic acids, coumarins and hydroxycinnamic acids) and isoprenoids, may account for these positive health effects, although the exact underlying mode of action is still unclear. Also, more studies considering the interplay of vitamins (C, E, folic acid), PUFAs and PPs are urgently required to provide insight into the nature of the effects of plant foods on brain aging and related maladies.

Acknowledgement

This work was supported by a research grant with the European Commission (EC), 5th Framework Program (QLK1-2001-00173).

References

1 Weinert BT, Timiras PS: Invited review: theories of aging. J Appl Physiol 2003;95:1706–1716.
2 Schmitt-Schillig S, Schaffer S, Weber CC, Eckert GP, Muller WE: Flavonoids and the aging brain. J Physiol Pharmacol 2005;56(suppl 1):23–36.

3 Dulak J: Nutraceuticals as antiangiogenic agents: hopes and reality. J Physiol Pharmacol 2005;56 (suppl 1):51–69.
4 Hancock JT, Desikan R, Neill SJ: Role of reactive oxygen species in cell signalling pathways. Biochem Soc Trans 2001;29:345–350.
5 Naseem KM: The role of nitric oxide in cardiovascular diseases. Mol Aspects Med 2005;26: 33–65.
6 Poon HF, Calabrese V, Scapagnini G, Butterfield DA: Free radicals and brain aging. Clin Geriatr Med 2004;20:329–359.
7 Sastre J, Pallardo FV, Vina J: The role of mitochondrial oxidative stress in aging. Free Radic Biol Med 2003;35:1–8.
8 Ballinger SW: Mitochondrial dysfunction in cardiovascular disease. Free Radic Biol Med 2005;38:1278–1295.
9 Singh KK, Luccy B, Zullo SJ: Mitochondria, oxidative stress and mitochondrial disease; in Cutler RG, Rodriguez H (eds): Critical Reviews of Oxidative Stress and Aging: Advances in Basic Science, Diagnostic and Intervention. Singapore, World Scientific Publishing, 2003.
10 Michikawa Y, Mazzucchelli F, Bresolin N, Scarlato G, Attardi G: Aging-dependent large accumulation of point mutations in the human mtDNA control region for replication. Science 1999;286: 774–779.
11 Bertoni-Freddari C, Fattoretti P, Giorgetti B, Solazzi M, Balietti M, Di Stefano G, Casoli T: Decay of mitochondrial metabolic competence in the aging cerebellum. Ann NY Acad Sci 2004;1019:29–32.
12 Judge S, Jang YM, Smith A, Hagen T, Leeuwenburgh C: Age-associated increases in oxidative stress and antioxidant enzyme activities in cardiac interfibrillar mitochondria: implications for the mitochondrial theory of aging. FASEB J 2005;19:419–421.
13 Cottrell DA, Turnbull DM: The mitochondrial genome, aging and neurodegenerative disorders; in Rosenberg IH, Sastre A (eds): Nutrition and Aging. Basel, Karger, 2002.
14 Numata T, Saito T, Maekawa K, Takahashi Y, Saitoh H, Hosokawa T, Fujita H, Kurasaki M: Bcl-2-linked apoptosis due to increase in NO synthase in brain of SAMP10. Biochem Biophys Res Commun 2002;297:517–522.
15 Manczak M, Jung Y, Park BS, Partovi D, Reddy PH: Time-course of mitochondrial gene expressions in mice brains: implications for mitochondrial dysfunction, oxidative damage, and cytochrome c in aging. J Neurochem 2005;92:494–504.
16 Brown-Borg HM, Harmann SM: Hormones and oxidative stress; in Cutler RG, Rodriguez H (eds): Critical Reviews of Oxidative Stress and Aging: Advances in Basic Science, Diagnostic and Intervention. Singapore, World Scientific Publishing, 2003.
17 Gluckman PD, Cutfield W, Hofman P, Hanson MA: The fetal, neonatal, and infant environments – the long-term consequences for disease risk. Early Hum Dev 2005;81:51–59.
18 Wu G, Bazer FW, Cudd TA, Meininger CJ, Spencer TE: Maternal nutrition and fetal development. J Nutr 2004;134:2169–2172.
19 Palinski W, Napoli C: The fetal origins of atherosclerosis: maternal hypercholesterolemia, and cholesterol-lowering or antioxidant treatment during pregnancy influence in utero programming and postnatal susceptibility to atherogenesis. FASEB J 2002;16:1348–1360.
20 Finch CE: Developmental origins of aging in brain and blood vessels: an overview. Neurobiol Aging 2005;26:281–291.
21 Reuter-Lorenz PA, Lustig C: Brain aging: reorganizing discoveries about the aging mind. Curr Opin Neurobiol 2005;15:245–251.
22 Fraser HB, Khaitovich P, Plotkin JB, Paabo S, Eisen MB: Aging and gene expression in the primate brain. PLoS Biol 2005;3:e274.
23 Toescu EC, Verkhratsky A, Landfield PW: Ca^{2+} regulation and gene expression in normal brain aging. Trends Neurosci 2004;27:614–620.
24 Muller WE, Hartmann H, Eckert A, Velbinger K, Forstl H: Free intracellular calcium in aging and Alzheimer's disease. Ann NY Acad Sci 1996;786:305–320.
25 Eddy DE, Harman D: Free radical theory of aging: effect of age, sex and dietary precursors on rat-brain docosahexanoic acid. J Am Geriatr Soc 1977;25:220–229.
26 Floyd RA: Antioxidants, oxidative stress, and degenerative neurological disorders. Proc Soc Exp Biol Med 1999;222:236–245.

27 Halliwell B: Reactive oxygen species and the central nervous system. J Neurochem 1992;59: 1609–1623.
28 Shukitt-Hale B, Youdim KA, Joseph JA: Neurological aging and nutritional intervention; in Cutler RG, Rodriguez H (eds): Critical Reviews of Oxidative Stress and Aging: Advances in Basic Science, Diagnostic and Intervention. Singapore, World Scientific Publishing, 2003.
29 Teoh CY, Davies KJA: Oxidative defense mechanisms; in Cutler RG, Rodriguez H (eds): Critical Reviews of Oxidative Stress and Aging: Advances in Basic Science, Diagnostic and Intervention. Singapore, World Scientific Publishing, 2003.
30 Marklund SL, Westman NG, Lundgren E, Roos G: Copper- and zinc-containing superoxide dismutase, manganese-containing superoxide dismutase, catalase, and glutathione peroxidase in normal and neoplastic human cell lines and normal human tissues. Cancer Res 1982;42: 1955–1961.
31 Reiter RJ: Oxidative processes and antioxidative defense mechanisms in the aging brain. FASEB J 1995;9:526–533.
32 Youdim KA, Shukitt-Hale B, Joseph JA: Flavonoids and the brain: interactions at the blood-brain barrier and their physiological effects on the central nervous system. Free Radic Biol Med 2004;37:1683–1693.
33 Shin CM, Chung YH, Kim MJ, Lee EY, Kim EG, Cha CI: Age-related changes in the distribution of nitrotyrosine in the cerebral cortex and hippocampus of rats. Brain Res 2002;931:194–199.
34 Leutner S, Eckert A, Muller WE: ROS generation, lipid peroxidation and antioxidant enzyme activities in the aging brain. J Neural Transm 2001;108:955–967.
35 Schuessel K, Schafer S, Bayer TA, Czech C, Pradier L, Muller-Spahn F, Muller WE, Eckert A: Impaired Cu/Zn-SOD activity contributes to increased oxidative damage in APP transgenic mice. Neurobiol Dis 2005;18:89–99.
36 Baek BS, Kwon HJ, Lee KH, Yoo MA, Kim KW, Ikeno Y, Yu BP, Chung HY: Regional difference of ROS generation, lipid peroxidation, and antioxidant enzyme activity in rat brain and their dietary modulation. Arch Pharm Res 1999;22:361–366.
37 Tsay HJ, Wang P, Wang SL, Ku HH: Age-associated changes of superoxide dismutase and catalase activities in the rat brain. J Biomed Sci 2000;7:466–474.
38 Mecocci P, Beal MF, Cecchetti R, Polidori MC, Cherubini A, Chionne F, Avellini L, Romano G, Senin U: Mitochondrial membrane fluidity and oxidative damage to mitochondrial DNA in aged and AD human brain. Mol Chem Neuropathol 1997;31:53–64.
39 Schuessel K, Leutner S, Cairns NJ, Muller WE, Eckert A: Impact of gender on upregulation of antioxidant defence mechanisms in Alzheimer's disease brain. J Neural Transm 2004;111: 1167–1182.
40 Smith DE, Rapp PR, McKay HM, Roberts JA, Tuszynski MH: Memory impairment in aged primates is associated with focal death of cortical neurons and atrophy of subcortical neurons. J Neurosci 2004;24:4373–4381.
41 Zhang A, Lorke DE, Lai HW, Chu X, Wu Y, Yew DT: Age-related alterations in cytochrome-c-mediated caspase activation in rhesus macaque monkey (*Macaca mulatta*) brains. Brain Res Mol Brain Res 2004;123:112–120.
42 Hu FB, Willett WC: Optimal diets for prevention of coronary heart disease. JAMA 2002;288: 2569–2578.
43 Genkinger JM, Platz EA, Hoffman SC, Comstock GW, Helzlsouer KJ: Fruit, vegetable, and antioxidant intake and all-cause, cancer, and cardiovascular disease mortality in a community-dwelling population in Washington County, Maryland. Am J Epidemiol 2004;160:1223–1233.
44 Willett WC, Sacks F, Trichopoulou A, Drescher G, Ferro-Luzzi A, Helsing E, Trichopoulos D: Mediterranean diet pyramid: a cultural model for healthy eating. Am J Clin Nutr 1995;61: 1402S–1406S.
45 Alonso A, de la FC, Martin-Arnau AM, de Irala J, Martinez JA, Martinez-Gonzalez MA: Fruit and vegetable consumption is inversely associated with blood pressure in a Mediterranean population with a high vegetable-fat intake: the Seguimiento Universidad de Navarra (SUN) Study. Br J Nutr 2004;92:311–319.
46 Heinrich M, Leonti M, Nebel S, Peschel W: 'Local Food-Nutraceuticals': an example of a multidisciplinary research project on local knowledge. J Physiol Pharmacol 2005;56(suppl 1):5–22.

47 Keys A: Mediterranean diet and public health: personal reflections. Am J Clin Nutr 1995;61: 1321S–1323S
48 Menotti A: Cardiovascular disease; in Spiller GA (ed): The Mediterranean Diets in Health and Disease. New York, AVI Books, 1991.
49 Ehnholm C, Huttunen JK, Pietinen P, Leino U, Mutanen M, Kostiainen E, Pikkarainen J, Dougherty R, Iacono J, Puska P: Effect of diet on serum lipoproteins in a population with a high risk of coronary heart disease. N Engl J Med 1982;307:850–855.
50 De Lorgeril M, Salen P, Martin JL, Monjaud I, Delaye J, Mamelle N: Mediterranean diet, traditional risk factors, and the rate of cardiovascular complications after myocardial infarction: final report of the Lyon Diet Heart Study. Circulation 1999;99:779–785.
51 Ambring A, Friberg P, Axelsen M, Laffrenzen M, Taskinen MR, Basu S, Johansson M: Effects of a Mediterranean-inspired diet on blood lipids, vascular function and oxidative stress in healthy subjects. Clin Sci (Lond) 2004;106:519–525.
52 Trichopoulou A, Orfanos P, Norat T, Bueno-de-Mesquita B, Ocke MC, Peeters PH, van der Schouw YT, Boeing H, Hoffmann K, Boffetta P, Nagel G, Masala G, Krogh V, Panico S, Tumino R, Vineis P, Bamia C, Naska A, Benetou V, Ferrari P, Slimani N, Pera G, Martinez-Garcia C, Navarro C, Rodriguez-Barranco M, Dorronsoro M, Spencer EA, Key TJ, Bingham S, Khaw KT, Kesse E, Clavel-Chapelon F, Boutron-Ruault MC, Berglund G, Wirfalt E, Hallmans G, Johansson I, Tjonneland A, Olsen A, Overvad K, Hundborg HH, Riboli E, Trichopoulos D: Modified Mediterranean diet and survival: EPIC-elderly prospective cohort study. BMJ 2005;330:991.
53 Harper CR, Jacobson TA: Beyond the Mediterranean diet: the role of omega–3 fatty acids in the prevention of coronary heart disease. Prev Cardiol 2003;6:136–146.
54 Simopoulos AP, Visioli F (eds): Mediterranean diets. World Rev Nutr Diet. Basel, Karger, 2000, vol 87.
55 Panza F, Solfrizzi V, Colacicco AM, D'Introno A, Capurso C, Torres F, Del Parigi A, Capurso S, Capurso A: Mediterranean diet and cognitive decline. Public Health Nutr 2004;7:959–963.
56 Yehuda S: Omega–6/omega–3 ratio and brain-related functions. World Rev Nutr Diet 2003;92: 37–56.
57 Alessandri JM, Guesnet P, Vancassel S, Astorg P, Denis I, Langelier B, Aid S, Poumes-Ballihaut C, Champeil-Potokar G, Lavialle M: Polyunsaturated fatty acids in the central nervous system: evolution of concepts and nutritional implications throughout life. Reprod Nutr Dev 2004;44: 509–538.
58 Bourre JM: Roles of unsaturated fatty acids (especially omega–3 fatty acids) in the brain at various ages and during ageing. J Nutr Health Aging 2004;8:163–174.
59 Haag M: Essential fatty acids and the brain. Can J Psychiatry 2003;48:195–203.
60 Simopoulos AP: Omega–3 fatty acids and antioxidants in edible wild plants. Biol Res 2004;37: 263–277.
61 Kris-Etherton PM, Zhao G, Binkoski AE, Coval SM, Etherton TD: The effects of nuts on coronary heart disease risk. Nutr Rev 2001;59:103–111.
62 Galli C, Przegalinsky E: Dietary induced long-lasting accumulation of 22:6 n–3 fatty acid in rat brain synaptosomal phospholipids. Pharmacol Res Commun 1973;5:239–248.
63 Moore SA: Polyunsaturated fatty acid synthesis and release by brain-derived cells in vitro. J Mol Neurosci 2001;16:195–200.
64 Jump DB: The biochemistry of n–3 polyunsaturated fatty acids. J Biol Chem 2002;277:8755–8758.
65 Yamagata K, Tagami M, Takenaga F, Yamori Y, Nara Y, Itoh S: Polyunsaturated fatty acids induce tight junctions to form in brain capillary endothelial cells. Neuroscience 2003;116:649–656.
66 Eckert GP, Wood WG, Muller WE: Effects of aging and β-amyloid on the properties of brain synaptic and mitochondrial membranes. J Neural Transm 2001;108:1051–1064.
67 Fernstrom JD: Effects of dietary polyunsaturated fatty acids on neuronal function. Lipids 1999;34:161–169.
68 Bourre JM, Bonneil M, Clement M, Dumont O, Durand G, Lafont H, Nalbone G, Piciotti M: Function of dietary polyunsaturated fatty acids in the nervous system. Prostaglandins Leukot Essent Fatty Acids 1993;48:5–15.
69 Clandinin MT, Jumpsen J, Suh MY: Relationship between fatty acid accretion, membrane composition, and biologic functions. J Pediatr 1994;125:S25–S32.

70 Yehuda S, Rabinovitz S, Carasso RL, Mostofsky DI: The role of polyunsaturated fatty acids in restoring the aging neuronal membrane. Neurobiol Aging 2002;23:843–853.
71 Kitajka K, Puskas LG, Zvara A, Hackler L, Barcelo-Coblijn G, Yeo YK, Farkas T: The role of n–3 polyunsaturated fatty acids in brain: modulation of rat brain gene expression by dietary n–3 fatty acids. Proc Natl Acad Sci USA 2002;99:2619–2624.
72 Kitajka K, Sinclair AJ, Weisinger RS, Weisinger HS, Mathai M, Jayasooriya AP, Halver JE, Puskas LG: Effects of dietary omega–3 polyunsaturated fatty acids on brain gene expression. Proc Natl Acad Sci USA 2004;101:10931–10936.
73 Puskas LG, Bereczki E, Santha M, Vigh L, Csanadi G, Spener F, Ferdinandy P, Onochy A, Kitajka K: Cholesterol and cholesterol plus DHA diet-induced gene expression and fatty acid changes in mouse eye and brain. Biochimie 2004;86:817–824.
74 Messeri GP, Oliverio A, Paoletti R, Galli C: Deficiency of essential fatty acids during pregnancy and avoidance learning in the progeny. Pharmacol Res Commun 1975;7:71–80.
75 Auestad N, Scott DT, Janowsky JS, Jacobsen C, Carroll RE, Montalto MB, Halter R, Qiu W, Jacobs JR, Connor WE, Connor SL, Taylor JA, Neuringer M, Fitzgerald KM, Hall RT: Visual, cognitive, and language assessments at 39 months: a follow-up study of children fed formulas containing long-chain polyunsaturated fatty acids to 1 year of age. Pediatrics 2003;112: e177–e183.
76 Richardson AJ, Puri BK: A randomized double-blind, placebo-controlled study of the effects of supplementation with highly unsaturated fatty acids on ADHD-related symptoms in children with specific learning difficulties. Prog Neuropsychopharmacol Biol Psychiatry 2002;26: 233–239.
77 Richardson AJ, Montgomery P: The Oxford-Durham study: a randomized, controlled trial of dietary supplementation with fatty acids in children with developmental coordination disorder. Pediatrics 2005;115:1360–1366.
78 McCann JC, Ames BN: Is docosahexaenoic acid, an n–3 long-chain polyunsaturated fatty acid, required for development of normal brain function? An overview of evidence from cognitive and behavioral tests in humans and animals. Am J Clin Nutr 2005;82:281–295.
79 Solfrizzi V, D'Introno A, Colacicco AM, Capurso C, Del Parigi A, Capurso S, Gadaleta A, Capurso A, Panza F: Dietary fatty acids intake: possible role in cognitive decline and dementia. Exp Gerontol 2005;40:257–270.
80 Suzuki H, Morikawa Y, Takahashi H: Effect of DHA oil supplementation on intelligence and visual acuity in the elderly. World Rev Nutr Diet 2001;88:68–71.
81 Visarius T, Azzi A: Alpha-tocopherol: beyond the antioxidant dogma; in Cutler RG, Rodriguez H (eds): Critical Reviews of Oxidative Stress and Aging: Advances in Basic Science, Diagnostic and Intervention. Singapore, World Scientific Publishing, 2003.
82 Rota C, Rimbach G, Minihane AM, Stoecklin E, Barella L: Dietary vitamin E modulates differential gene expression in the rat hippocampus: potential implications for its neuroprotective properties. Nutr Neurosci 2005;8:21–29.
83 Azzi A, Gysin R, Kempna P, Munteanu A, Negis Y, Villacorta L, Visarius T, Zingg JM: Vitamin E mediates cell signaling and regulation of gene expression. Ann NY Acad Sci 2004;1031:86–95.
84 Ricciarelli R, Zingg JM, Azzi A: Vitamin E: protective role of a Janus molecule. FASEB J 2001;15:2314–2325.
85 Gaziano JM: Vitamin E and cardiovascular disease: observational studies. Ann NY Acad Sci 2004;1031:280–291.
86 Fleshner NE: Vitamin E and prostate cancer. Urol Clin North Am 2002;29:107–113, ix.
87 Shekelle PG, Morton SC, Jungvig LK, Udani J, Spar M, Tu W, Suttorp J, Coulter I, Newberry SJ, Hardy M: Effect of supplemental vitamin E for the prevention and treatment of cardiovascular disease. J Gen Intern Med 2004;19:380–389.
88 Eidelman RS, Hollar D, Hebert PR, Lamas GA, Hennekens CH: Randomized trials of vitamin E in the treatment and prevention of cardiovascular disease. Arch Intern Med 2004;164:1552–1556.
89 Sano M, Ernesto C, Thomas RG, Klauber MR, Schafer K, Grundman M, Woodbury P, Growdon J, Cotman CW, Pfeiffer E, Schneider LS, Thal LJ: A controlled trial of selegiline, α-tocopherol, or both as treatment for Alzheimer's disease. The Alzheimer's Disease Cooperative Study. N Engl J Med 1997;336:1216–1222.

90 Petersen RC, Thomas RG, Grundman M, Bennett D, Doody R, Ferris S, Galasko D, Jin S, Kaye J, Levey A, Pfeiffer E, Sano M, van Dyck CH, Thal LJ: Vitamin E and donepezil for the treatment of mild cognitive impairment. N Engl J Med 2005;352:2379–2388.
91 Navarro A, Gomez C, Sanchez-Pino MJ, Gonzalez H, Bandez MJ, Boveris AD, Boveris A: Vitamin E at high doses improves survival, neurological performance and brain mitochondrial function in aging male mice. Am J Physiol Regul Integr Comp Physiol 2005;289:R1392–R1399.
92 Sumien N, Heinrich KR, Sohal RS, Forster MJ: Short-term vitamin E intake fails to improve cognitive or psychomotor performance of aged mice. Free Radic Biol Med 2004;36: 1424–1433.
93 Jiang Q, Christen S, Shigenaga MK, Ames BN: Gamma-tocopherol, the neglected form of vitamin E: new functions, new interest; in Cutler RG, Rodriguez H (eds): Critical Reviews of Oxidative Stress and Aging: Advances in Basic Science, Diagnostic and Intervention. Singapore, World Scientific Publishing, 2003.
94 Jiang Q, Ames BN: Gamma-tocopherol, but not α-tocopherol, decreases proinflammatory eicosanoids and inflammation damage in rats. FASEB J 2003;17:816–822.
95 Wagner KH, Kamal-Eldin A, Elmadfa I: Gamma-tocopherol – an underestimated vitamin? Ann Nutr Metab 2004;48:169–188.
96 Vatassery GT, Adityanjee, Quach HT, Smith WE, Kuskowski MA, Melnyk D: Alpha- and γ-tocopherols in cerebrospinal fluid and serum from older, male, human subjects. J Am Coll Nutr 2004;23:233–238.
97 Morris MC, Evans DA, Tangney CC, Bienias JL, Wilson RS, Aggarwal NT, Scherr PA: Relation of the tocopherol forms to incident Alzheimer disease and to cognitive change. Am J Clin Nutr 2005;81:508–514.
98 Schaffer S, Muller WE, Eckert GP: Tocotrienols: constitutional effects in aging and disease. J Nutr 2005;135:151–154.
99 Sen CK, Khanna S, Roy S: Tocotrienol: the natural vitamin E to defend the nervous system? Ann NY Acad Sci 2004;1031:127–142.
100 Grunewald RA: Ascorbic acid in the brain. Brain Res Brain Res Rev 1993;18:123–133.
101 Naidu KA: Vitamin C in human health and disease is still a mystery? An overview. Nutr J 2003;2:7.
102 Van der LB, Bachschmid M, Spitzer V, Brey L, Ullrich V, Luscher TF: Decreased plasma and tissue levels of vitamin C in a rat model of aging: implications for antioxidative defense. Biochem Biophys Res Commun 2003;303:483–487.
103 Wilson JX: Regulation of vitamin C transport. Annu Rev Nutr 2005;25:105–125.
104 Riviere S, Birlouez-Aragon I, Nourhashemi F, Vellas B: Low plasma vitamin C in Alzheimer patients despite an adequate diet. Int J Geriatr Psychiatry 1998;13:749–754.
105 Huang J, Agus DB, Winfree CJ, Kiss S, Mack WJ, McTaggart RA, Choudhri TF, Kim LJ, Mocco J, Pinsky DJ, Fox WD, Israel RJ, Boyd TA, Golde DW, Connolly ES Jr: Dehydroascorbic acid, a blood-brain barrier transportable form of vitamin C, mediates potent cerebroprotection in experimental stroke. Proc Natl Acad Sci USA 2001;98:11720–11724.
106 Rice ME: Ascorbate regulation and its neuroprotective role in the brain. Trends Neurosci 2000;23:209–216.
107 Lunec J: Ascorbic acid DNA damage and repair; in Cutler RG, Rodriguez H (eds): Critical Reviews of Oxidative Stress and Aging: Advances in Basic Science, Diagnostic and Intervention. Singapore, World Scientific Publishing, 2003.
108 Masaki KH, Losonczy KG, Izmirlian G, Foley DJ, Ross GW, Petrovitch H, Havlik R, White LR: Association of vitamin E and C supplement use with cognitive function and dementia in elderly men. Neurology 2000;54:1265–1272.
109 Paleologos M, Cumming RG, Lazarus R: Cohort study of vitamin C intake and cognitive impairment. Am J Epidemiol 1998;148:45–50.
110 Grodstein F, Chen J, Willett WC: High-dose antioxidant supplements and cognitive function in community-dwelling elderly women. Am J Clin Nutr 2003;77:975–984.
111 Mendelsohn AB, Belle SH, Stoehr GP, Ganguli M: Use of antioxidant supplements and its association with cognitive function in a rural elderly cohort: the MoVIES Project. Monongahela Valley Independent Elders Survey. Am J Epidemiol 1998;148:38–44.

112 Maxwell CJ, Hicks MS, Hogan DB, Basran J, Ebly EM: Supplemental use of antioxidant vitamins and subsequent risk of cognitive decline and dementia. Dement Geriatr Cogn Disord 2005;20:45–51.
113 Sun AY, Simonyi A, Sun GY: The 'French paradox' and beyond: neuroprotective effects of polyphenols. Free Radic Biol Med 2002;32:314–318.
114 Heim KE, Tagliaferro AR, Bobilya DJ: Flavonoid antioxidants: chemistry, metabolism and structure-activity relationships. J Nutr Biochem 2002;13:572–584.
115 Harborne JB, Williams CA: Advances in flavonoid research since 1992. Phytochemistry 2000;55:481–504.
116 Frank J: Beyond vitamin E supplementation: an alternative strategy to improve vitamin E status. J Plant Physiol 2005;162:834–843.
117 Janssen K, Mensink RP, Cox FJ, Harryvan JL, Hovenier R, Hollman PC, Katan MB: Effects of the flavonoids quercetin and apigenin on hemostasis in healthy volunteers: results from an in vitro and a dietary supplement study. Am J Clin Nutr 1998;67:255–262.
118 Radtke J, Linseisen J, Wolfram G: Fasting plasma concentrations of selected flavonoids as markers of their ordinary dietary intake. Eur J Nutr 2002;41:203–209.
119 Engler MB, Engler MM, Chen CY, Malloy MJ, Browne A, Chiu EY, Kwak HK, Milbury P, Paul SM, Blumberg J, Mietus-Snyder ML: Flavonoid-rich dark chocolate improves endothelial function and increases plasma epicatechin concentrations in healthy adults. J Am Coll Nutr 2004;23:197–204.
120 Carbonaro M, Grant G: Absorption of quercetin and rutin in rat small intestine. Ann Nutr Metab 2005;49:178–182.
121 Baba S, Osakabe N, Natsume M, Muto Y, Takizawa T, Terao J: In vivo comparison of the bioavailability of (+)-catechin, (−)-epicatechin and their mixture in orally administered rats. J Nutr 2001;131:2885–2891.
122 Henning SM, Niu Y, Liu Y, Lee NH, Hara Y, Thames GD, Minutti RR, Carpenter CL, Wang H, Heber D: Bioavailability and antioxidant effect of epigallocatechin gallate administered in purified form versus as green tea extract in healthy individuals. J Nutr Biochem 2005;16:610–616.
123 Zern TL, Fernandez ML: Cardioprotective effects of dietary polyphenols. J Nutr 2005;135:2291–2294.
124 Commenges D, Scotet V, Renaud S, Jacqmin-Gadda H, Barberger-Gateau P, Dartigues JF: Intake of flavonoids and risk of dementia. Eur J Epidemiol 2000;16:357–363.
125 Benzie IF: Evolution of dietary antioxidants. Comp Biochem Physiol A Mol Integr Physiol 2003;136:113–126.
126 Schindowski K, Leutner S, Kressmann S, Eckert A, Muller WE: Age-related increase of oxidative stress-induced apoptosis in mice prevention by *Ginkgo biloba* extract (EGb761). J Neural Transm 2001;108:969–978.
127 Stoll S, Scheuer K, Pohl O, Muller WE: *Ginkgo biloba* extract (EGb 761) independently improves changes in passive avoidance learning and brain membrane fluidity in the aging mouse. Pharmacopsychiatry 1996;29:144–149.
128 Lee L, Kang SA, Lee HO, Lee BH, Park JS, Kim JH, Jung IK, Park YJ, Lee JE: Relationships between dietary intake and cognitive function level in Korean elderly people. Public Health 2001;115:133–138.
129 Kang JH, Ascherio A, Grodstein F: Fruit and vegetable consumption and cognitive decline in aging women. Ann Neurol 2005;57:713–720.
130 Zeghichi S, Kallithraka S, Simopoulos A, Kypriotakis Z: Nutritional composition of selected wild plants in the diet of Crete. World Rev Nutr Diet 2003;91:22–40.
131 Trichopoulou A, Vasilopoulou E, Hollman P, Chamalides CH, Foufa E, Kaloudis TR, Kromhout D, Miskaki PH, Petrochilou I, Poulima E, Stafilakis K, Theophilou D: Nutritional composition and flavonoid content of edible wild greens and green pies: a potential rich source of antioxidant nutrients in the Mediterranean diet. Food Chem 2000;70:319–323.
132 Visioli F, Galli C: The role of antioxidants in the Mediterranean diet. Lipids 2001;36(suppl):S49–S52.
133 The Local-Food Nutraceuticals Consortium: Understanding local Mediterranean diets: A multidisciplinary pharmacological and ethnobotanical approach. Pharmacol Res 2005;52:353–366.

134 Schaffer S, Schmitt-Schillig S, Muller WE, Eckert GP: Antioxidant properties of Mediterranean food plant extracts: geographical differences. J Physiol Pharmacol 2005;56(suppl 1):115–124.
135 Schroeter H, Spencer JEA: Flavonoids: neuroprotective agents? Modulation of oxidative stress-induced MAP kinase signal transduction; in Rice-Evans CA, Packer L (eds): Flavonoids in Health and Disease. New York, Dekker, 2003.
136 Mandel SA, Avramovich-Tirosh Y, Reznichenko L, Zheng H, Weinreb O, Amit T, Youdim MB: Multifunctional activities of green tea catechins in neuroprotection. Modulation of cell survival genes, iron-dependent oxidative stress and PKC signaling pathway. Neurosignals 2005;14:46–60.
137 Dore S: Unique properties of polyphenol stilbenes in the brain: more than direct antioxidant actions; gene/protein regulatory activity. Neurosignals 2005;14:61–70.
138 Rimbach G, Wolffram S, Watanabe C, Packer L, Gohil K: Effect of *Ginkgo biloba* (EGb 761) on differential gene expression. Pharmacopsychiatry 2003;36(suppl 1):S95–S99.
139 Hui KM, Huen MS, Wang HY, Zheng H, Sigel E, Baur R, Ren H, Li ZW, Wong JT, Xue H: Anxiolytic effect of wogonin, a benzodiazepine receptor ligand isolated from *Scutellaria baicalensis* Georgi. Biochem Pharmacol 2002;64:1415–1424.
140 Dekermendjian K, Kahnberg P, Witt MR, Sterner O, Nielsen M, Liljefors T: Structure-activity relationships and molecular modeling analysis of flavonoids binding to the benzodiazepine site of the rat brain $GABA_A$ receptor complex. J Med Chem 1999;42:4343–4350.
141 Simopoulos AP, Norman HA, Gillaspy JE: Purslane in human nutrition and its potential for world agriculture. World Rev Nutr Diet 1995;77:47–74.
142 Radhakrishnan R, Zakaria MN, Islam MW, Chen HB, Kamil M, Chan K, Al Attas A: Neuropharmacological actions of *Portulaca oleraceae* L. var. *sativa* (Hawk). J Ethnopharmacol 2001;76:171–176.
143 Newman AB, Fitzpatrick AL, Lopez O, Jackson S, Lyketsos C, Jagust W, Ives D, Dekosky ST, Kuller LH: Dementia and Alzheimer's disease incidence in relationship to cardiovascular disease in the cardiovascular health study cohort. J Am Geriatr Soc 2005;53:1101–1107.
144 Ravona-Springer R, Davidson M, Noy S: The role of cardiovascular risk factors in Alzheimer's disease. CNS Spectr 2003;8:824–833.
145 Schaffer S, Eckert GP, Muller WE, Llorach R, Rivera D, Grande S, Galli C, Visioli F: Hypochlorous acid scavenging properties of local Mediterranean plant foods. Lipids 2004;39:1239–1247.
146 Cao G, Russell RM, Lischner N, Prior RL: Serum antioxidant capacity is increased by consumption of strawberries, spinach, red wine or vitamin C in elderly women. J Nutr 1998;128:2383–2390.
147 Lampe JW: Health effects of vegetables and fruit: assessing mechanisms of action in human experimental studies. Am J Clin Nutr 1999;70:475S-490S.
148 Caruso D, Rise P, Galella G, Regazzoni C, Toia A, Galli G, Galli C: Formation of 22 and 24 carbon 6-desaturated fatty acids from exogenous deuterated arachidonic acid is activated in THP-1 cells at high substrate concentrations. FEBS Lett 1994;343:195–199.
149 Sprecher H: Metabolism of highly unsaturated n–3 and n–6 fatty acids. Biochim Biophys Acta 2000;1486:219–231.
150 Patil CS, Singh VP, Satyanarayan PS, Jain NK, Singh A, Kulkarni SK: Protective effect of flavonoids against aging- and lipopolysaccharide-induced cognitive impairment in mice. Pharmacology 2003;69:59–67.
151 Thiyagarajan M, Sharma SS: Neuroprotective effect of curcumin in middle cerebral artery occlusion induced focal cerebral ischemia in rats. Life Sci 2004;74:969–985.
152 Lee SR, Im KJ, Suh SI, Jung JG: Protective effect of green tea polyphenol (–)-epigallocatechin gallate and other antioxidants on lipid peroxidation in gerbil brain homogenates. Phytother Res 2003;17:206–209.
153 Choi YB, Kim YI, Lee KS, Kim BS, Kim DJ: Protective effect of epigallocatechin gallate on brain damage after transient middle cerebral artery occlusion in rats. Brain Res 2004;1019:47–54.
154 Heo HJ, Kim MJ, Lee JM, Choi SJ, Cho HY, Hong B, Kim HK, Kim E, Shin DH: Naringenin from *Citrus junos* has an inhibitory effect on acetylcholinesterase and a mitigating effect on amnesia. Dement Geriatr Cogn Disord 2004;17:151–157.
155 Abd El-Gawad HM, Khalifa AE: Quercetin, coenzyme Q10, and *L*-canavanine as protective agents against lipid peroxidation and nitric oxide generation in endotoxin-induced shock in rat brain. Pharmacol Res 2001;43:257–263.

156 Wang Q, Yu S, Simonyi A, Rottinghaus G, Sun GY, Sun AY: Resveratrol protects against neurotoxicity induced by kainic acid. Neurochem Res 2004;29:2105–2112.
157 Sharma M, Gupta YK: Chronic treatment with trans resveratrol prevents intracerebroventricular streptozotocin-induced cognitive impairment and oxidative stress in rats. Life Sci 2002;71:2489–2498.
158 Florek E, Ignatowicz E, Wrzosek J, Piekoszewski W: Effect of rutin on total antioxidant status of rats exposed to cigarette smoke. Pharmacol Rep 2005;57:84–89.
159 Cho J, Lee HK: Wogonin inhibits ischemic brain injury in a rat model of permanent middle cerebral artery occlusion. Biol Pharm Bull 2004;27:1561–1564.
160 Parle M, Dhingra D: Ascorbic acid: a promising memory-enhancer in mice. J Pharmacol Sci 2003;93:129–135.
161 Wozniak A, Wozniak B, Drewa G, Kasprzak HA: The effect of ascorbic acid and ethanol on the level of thiobarbituric acid reactive substances in selected tissues of albino BALB/c mice. Biol Pharm Bull 2002;25:943–944.
162 Zaidi SM, Banu N: Antioxidant potential of vitamins A, E and C in modulating oxidative stress in rat brain. Clin Chim Acta 2004;340:229–233.
163 Henry PT, Chandy MJ: Effect of ascorbic acid on infarct size in experimental focal cerebral ischaemia and reperfusion in a primate model. Acta Neurochir (Wien) 1998;140:977–980.
164 Devi SA, Kiran TR: Regional responses in antioxidant system to exercise training and dietary vitamin E in aging rat brain. Neurobiol Aging 2004;25:501–508.
165 Wu A, Ying Z, Gomez-Pinilla F: The interplay between oxidative stress and brain-derived neurotrophic factor modulates the outcome of a saturated fat diet on synaptic plasticity and cognition. Eur J Neurosci 2004;19:1699–1707.
166 Zararsiz I, Kus I, Akpolat N, Songur A, Ogeturk M, Sarsilmaz M: Protective effects of omega–3 essential fatty acids against formaldehyde-induced neuronal damage in prefrontal cortex of rats. Cell Biochem Funct 2005;24:237–244.
167 Lim GP, Calon F, Morihara T, Yang FS, Teter B, Ubeda O, Salem N, Frautschy SA, Cole GM: A diet enriched with the omega–3 fatty acid docosahexaenoic acid reduces amyloid burden in an aged Alzheimer mouse model. J Neurosci 2005;25:3032–3040.
168 Choi-Kwon S, Park KA, Lee HJ, Park MS, Lee JH, Jeon SE, Choe MA, Park KC: Temporal changes in cerebral antioxidant enzyme activities after ischemia and reperfusion in a rat focal brain ischemia model: effect of dietary fish oil. Brain Res Dev Brain Res 2004;152:11–18.
169 Talavera S, Felgines C, Texier O, Besson C, Gil-Izquierdo A, Lamaison JL, Remesy C: Anthocyanin metabolism in rats and their distribution to digestive area, kidney, and brain. J Agric Food Chem 2005;53:3902–3908.
170 De Boer VC, Dihal AA, van der WH, Arts IC, Wolffram S, Alink GM, Rietjens IM, Keijer J, Hollman PC: Tissue distribution of quercetin in rats and pigs. J Nutr 2005;135:1718–1725.
171 Andres-Lacueva C, Shukitt-Hale B, Galli RL, Jauregui O, Lamuela-Raventos RM, Joseph JA: Anthocyanins in aged blueberry-fed rats are found centrally and may enhance memory. Nutr Neurosci 2005;8:111–120.
172 Wang Y, Chang CF, Chou J, Chen HL, Deng X, Harvey BK, Cadet JL, Bickford PC: Dietary supplementation with blueberries, spinach, or spirulina reduces ischemic brain damage. Exp Neurol 2005;193:75–84.
173 Joseph JA, Shukitt-Hale B, Denisova NA, Bielinski D, Martin A, McEwen JJ, Bickford PC: Reversals of age-related declines in neuronal signal transduction, cognitive, and motor behavioral deficits with blueberry, spinach, or strawberry dietary supplementation. J Neurosci 1999;19: 8114–8121.
174 Bickford PC, Gould T, Briederick L, Chadman K, Pollock A, Young D, Shukitt-Hale B, Joseph J: Antioxidant-rich diets improve cerebellar physiology and motor learning in aged rats. Brain Res 2000;866:211–217.
175 Nishiyama N, Moriguchi T, Saito H: Beneficial effects of aged garlic extract on learning and memory impairment in the senescence-accelerated mouse. Exp Gerontol 1997;32:149–160.
176 Hwang IK, Yoo KY, Kim DS, Jeong YK, Kim JD, Shin HK, Lim SS, Yoo ID, Kang TC, Kim DW, Moon WK, Won MH: Neuroprotective effects of grape seed extract on neuronal injury by inhibiting DNA damage in the gerbil hippocampus after transient forebrain ischemia. Life Sci 2004;75:1989–2001.

177 Balu M, Sangeetha P, Murali G, Panneerselvam C: Age-related oxidative protein damages in central nervous system of rats: modulatory role of grape seed extract. Int J Dev Neurosci 2005;23: 501–507.
178 Balu M, Sangeetha P, Haripriya D, Panneerselvam C: Rejuvenation of antioxidant system in central nervous system of aged rats by grape seed extract. Neurosci Lett 2005;383:295–300.
179 Unno K, Takabayashi F, Kishido T, Oku N: Suppressive effect of green tea catechins on morphologic and functional regression of the brain in aged mice with accelerated senescence (SAMP10). Exp Gerontol 2004;39:1027–1034.
180 Hong JT, Ryu SR, Kim HJ, Lee JK, Lee SH, Yun YP, Lee BM, Kim PY: Protective effect of green tea extract on ischemia/reperfusion-induced brain injury in Mongolian gerbils. Brain Res 2001;888:11–18.
181 Kim HK, Kim M, Kim S, Kim M, Chung JH: Effects of green tea polyphenol on cognitive and acetylcholinesterase activities. Biosci Biotechnol Biochem 2004;68:1977–1979.
182 Scarmeas N, Stern Y, Tang M-X, Mayeux R, Luchsinger JA: Mediterranean diet and risk for Alzheimer's disease. Ann Neurol DOI: 10.1002/ana.20854 (accessed April 25, 2006).

Prof. Dr. Walter E. Müller
Institute of Pharmacology (ZAFES), Biocenter Niederursel
University of Frankfurt, Max-von-Laue-Strasse 9
DE–60438 Frankfurt (Germany)
Tel. +49 69 7982 9376, Fax +49 69 7982 9374, E-Mail PharmacolNat@em.uni-frankfurt.de

Heinrich M, Müller WE, Galli C (eds): Local Mediterranean Food Plants and Nutraceuticals. Forum Nutr. Basel, Karger, 2006, vol 59, pp 116–129

Local Food and Cardioprotection: The Role of Phytochemicals

Francesco Visioli[a], *Paola Bogani*[a], *Simona Grande*[a], *Vivian Detopoulou*[b], *Yannis Manios*[b], *Claudio Galli*[a]

[a]Department of Pharmacological Sciences, University of Milan, Milan, Italy;
[b]Department of Nutrition and Dietetics, Harokopio University, Athens, Greece

Abstract

For decades, most of the attention of nutritionists and health professionals has focused on the impact of the major dietary components, such as the amounts and types of fats, proteins, carbohydrates and fibers, on human health. However, interest in the role of minor components is rapidly growing. Many constituents of plants are non-nutritional compounds that play key roles in plant physiology and interactions with the environment. Over the past few years, we performed human studies to ascertain the health effects of Mediterranean foods such as extra virgin olive oil and tomatoes. Recently, we became interested in endothelial dysfunction and its implications in aging. To study the effects of local food plants on vascular function, plants were collected in Southern Italy. Extracts were first tested for their antioxidant activity in a variety of assays. The effects on the production of vasorelaxant factors were then investigated in cell culture. Finally, aged rats were fed with a wild artichoke extract and their vasomotion responsiveness was evaluated. In synthesis, the data uniformly demonstrate that phytochemical components of the Mediterranean diet exert cardioprotective effects whose mechanisms are being progressively elucidated.

Local Dietary Habits: Past and Present

The human diet has been largely based, until recent times, upon the consumption of natural foods, from animals and plants. Natural foods are not just a mixture of chemicals, but are made of parts (organs, tissues, and cells) of other organisms, either plants or animals. Hunter-gatherers from the beginning of history until recent times, when agriculture was initiated and developed, and unprocessed food items, directly available in the environment and consumed immediately or stored for short periods, represented the bases of nutrition. The development of agriculture, based on cultivation of edible plants, in addition to animal breeding has allowed humans to become sedentary and to increase the global population. However, consumption of wild plants has remained part of local dietary traditions and habits and is still followed by small population groups, generally elderly subjects, in local communities. This situation has been especially maintained in rural communities living in remote areas of the Mediterranean basin: these habits may have contributed to the healthful effects of the typical Mediterranean diets. Consumption of certain plants or of selected parts of them is still part of folk medicine and is used in the treatment of certain diseases. Although most of this 'local culture' has not been subjected to adequate controlled investigations and to scientific criticism, the concept that micronutrients contribute to the protective effects of these diets is commonly accepted [1].

Nutrition and Human Health: From Physiology to Pathology

For decades, most of the attention of nutritionists and health professionals has focused on the impact of the major dietary components, such as the amounts and types of fats, proteins, carbohydrates and fibers, on human health. However, interest in the role of minor components is rapidly growing. In fact, the presence of vitamins, provitamins, and a variety of antioxidants from flavonoids, catechins and anthocyanins to simple phenols in a diet rich in edible plants is certainly relevant. Many constituents of plants are non-nutritional compounds that play key roles in plant physiology and interactions with the environment [2]. More recently, such constituents have been shown to exert protective effects in humans, as they play a role in the antioxidant defense of the organism. Moreover, in addition to their mere antioxidant actions, their effects on various enzyme activities appear to be even more promising: research in this area should be further extensively promoted. Some of the difficulties in such studies, however, are related to the complexity of isolating the effects of numerous biofactors present in complex dietcs, e.g. the various compositions [3–8] of the foods themselves, especially because of the possible synergy between the various biofactors [9].

It is part of the research strategy in this area to identify, isolate, and test individual compounds or mixtures of them for their effects on biological systems, from cellular to animal and, possibly, human studies. Additional relevant issues in research strategies in this area are: assessment of their bioavailability, which is dependent not only upon their chemistry, but also upon the matrix in which they are administered (food vs. formulations), and their associations [10]; evaluation of the correlations between plasma levels and biological effects, through the development of appropriate relevant biomarkers [11].

The chemistry, classification, and sources of potentially bioactive compounds and their consumption by various population groups represent a complex area of investigation, beyond the scope of this chapter, and the reader is referred to other chapters of this volume and to other authors [12] for this type of information.

The concept that nutrition plays a major role in controlling cell, organ, and system function is supported by the impressive alteration of the health status of large population groups observed over the last decades in relation to modifications in lifestyles and dietary habits. The rates of incidence of diseases or syndromes leading to pathologies are rapidly growing not only in Western nations but also in significant sections of developing countries [13]. These diseases affect the cardiovascular system (atherosclerosis) or lead to metabolic disorders (metabolic syndrome and associated symptoms, i.e. type 2 diabetes, obesity, hypertension, alterations of lipid and lipoprotein metabolism) and also cancer [14]. Although genetic predisposition might expose selected individuals, this is certainly not the case for large populations. In fact, the rapid increments in the incidence of the above-mentioned disorders are related to major changes in lifestyle, namely dietary habits that took place in a few decades in large population groups and cannot be attributed to genetic changes. To quote M.P. Stern: 'Genetics loads the gun, but the environment pulls the trigger'.

Studies on the correlations between dietary and lifestyle factors and health have rapidly pointed the attention to the impact of diet in the area CVD, due to the continuous exposure of vessel walls to components derived from the diet. Evidence of the 'protective' effects of a diet, namely the Mediterranean diets and particularly the diet of Crete, has been produced by several studies [15–21]. However, this area deserves further investigation, also because of the number of factors involved. Moreover, accumulating evidence suggests that the incidence of 'classic' risk factors for CHD, such as blood cholesterol and triglycerides, is not much different between populations of the Mediterranean region and those of other areas such as Northern Europe and the USA [22, 23].

More recently, attention to the relationships between the diet and vascular function, including metabolic parameters, e.g. glycemia and insulin responses, blood pressure, etc., is being focused on the major functional and metabolic changes that occur during the postprandial state, a condition that lasts for several hours/days in the human [24]. This interest has been sparked by recent evidence that the endothelium is an important barrier that protects against the development of atherosclerosis. Endothelial dysfunction is a major complication of atherosclerosis [25, 26] and its incidence increases with age. It is now clear that oxidative stress plays a significant role in the onset and development of endothelial dysfunction. These effects are mediated by a reduced production/availability of the vasorelaxant factor nitric oxide (NO) [27–29]. Accordingly, administration of antioxidants, e.g. vitamin C and flavonoids from tea and wine, has been shown to ameliorate endothelial function and restore proper vasomotion [29, 30], likely as a consequence of increased endothelial NO synthase (eNOS) activity.

This chapter focuses on the healthful and cardioprotective effects of compounds that are typically present in Mediterranean foods, such as extra virgin olive oil, tomatoes and, as part of the 'Local Food-Nutraceuticals' project, of compounds extracted from wild plants which are part of traditional diets.

Local Food Plants and Vascular Function

To study the effects of local food plants on vascular function, plants were collected in Southern Italy (Castelmezzano) and were dried and extracted with 90% hot ethanol under reflux. The total polyphenolic content of the extracts was determined by the Folin-Ciocalteau method, using gallic acid as the reference compound [31]. The extracts were analyzed by HPLC coupled with ESI and DAD-UV detectors, using a Waters Xterra MS-C18 column and a H_2O:AcN/MeOH gradient elution. Major peaks were identified by comparison with authentic standards and a database.

The potential to induce vasorelaxation was studied in vitro by the assessment of the production of NO, both in porcine aortic endothelial cells (PAEC) and in brain cell homogenates, where the use of an NO sensor made it possible to evaluate directly the release of NO. Furthermore, the production of another vasorelaxant factor, i.e. prostacyclin, assessed as the formation of its stable metabolite 6-keto-$PGF_{1\alpha}$, was evaluated in PAEC [32].

The antioxidant activities of plant food extracts were assessed by using hypochlorous acid as the toxic agent, as its overproduction has been implicated in the onset of atherosclerosis and endothelial dysfunction. Thus, we studied the HOCl-induced modifications of apoproteins, namely of apoB, in isolated LDL [33].

Supplementation of PAEC with *Cynara cardunculus* or *Thymus pulegioides* extracts (10^{-6} mol/l gallic acid equivalents) significantly increased eNOS activity by 234 and 135%, respectively, as assessed by increased conversion of *L*-arginine to *L*-citrulline table 1. This effect was inhibited by co-incubation with *L*-NNA 300 μM, further suggesting that the tonic effects of plant extracts are due to enhanced enzymatic activity (table 1).

The production of prostacyclin was also significantly increased (+269 and +190%, respectively) by *C. cardunculus* and *T. pulegioides* supplementation (table 1).

Table 1. Effect of local food plant extracts on the production of vasorelaxant factors (NO and prostacyclin) by endothelial cells

	eNOS activity, pmol citrulline/mg protein	+*L*-NNA	$PGF_{1\alpha}$ μg/mg protein
Control	13.4 ± 1.3	13.4 ± 1.3	138.3 ± 5.3
A23187	37.5 ± 4.5	37.5 ± 4.5	170.6 ± 8.6
L-NNA		15.4 ± 2.9	
LYNC 10^{-6} *M*	83.5 ± 5.6*	18 ± 5.6	372.3 ± 9.8*
LTHY 10^{-6} *M*	50.6 ± 4.3*	20.5 ± 4.3	262.6 ± 7.6*

Porcine aortic endothelial cells (PAEC) were incubated with *C. cardunculus* or *T. pulegioides* extracts to determine prostacyclin production and eNOS activity either in the absence or in the presence of the eNOS inhibitor *L*-NNA (*N*-nitro-*L*-arginine, 300 μmol/l). Prostacyclin production was assessed as the formation of its stable metabolite 6-keto-$PGF_{1\alpha}$. Data are expressed as means ± SD, $^*p < 0.05$. From Grande et al. [32].

Enhanced NO production by *C. cardunculus* addition was further verified (in a rat brain homogenate) by a direct method, that is by the use of a sensitive NO sensor. Co-incubation with cardoon extract 10^{-5} *M* increased NO production by 35.4%, i.e. from 56.2 ± 3.6 to 75.6 ± 4.1 nmol/l peak values over a 5-min time period figure 1. This effect was also inhibited by *L*-NNA 300 μmol/l.

Western blot analyses of cell homogenates and RT-PCR experiments demonstrated that eNOS expression was not altered by either extract, indicating that the stimulating effects on NO production were due to a direct activity on eNOS activity rather than synthesis [32].

It is noteworthy that recent studies using isolated rat vessels also showed that *C. cardunculus* exerts vasorelaxant effects. Most important, feeding studies confirmed that a *C. cardunculus* extract can restore proper vascular function in aged Fisher 344 × BN rats fed the extract (10 mg polyphenols/kg) for 5 days [34]. These data confirm in vivo the results we observed in vitro and build the grounds for the formulation of nutraceuticals aimed at maintaining vascular motility in the elderly.

Protection from HOCl-Induced Amino Acid Modifications

Amino acidic residues of apoB are susceptible to oxidation by HOCl, which leads to their uncontrolled uptake by macrophages [35]. As shown in figure 2, addition of HOCl to LDL samples caused the modification of apoB, in particular

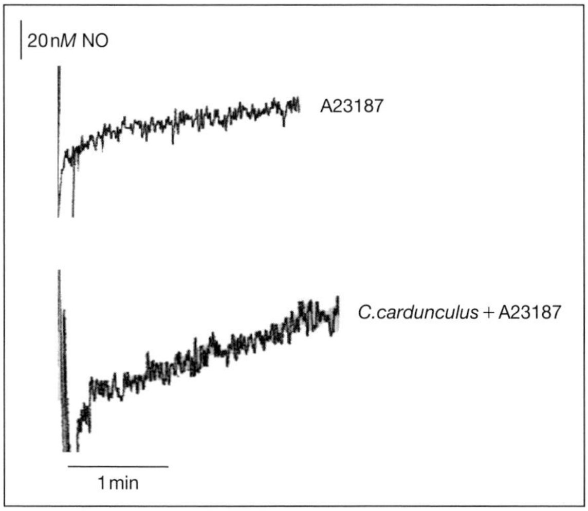

Fig. 1. Real-time representation of the effect of *C. cardunculus* (10^{-5} M) on NO production by brain cell membranes. NO production was stimulated by the addition of the calcium ionophore A23187 (2 μmol/l) and was continually recorded with an NO-specific amperometric electrode. From Grande et al. [32].

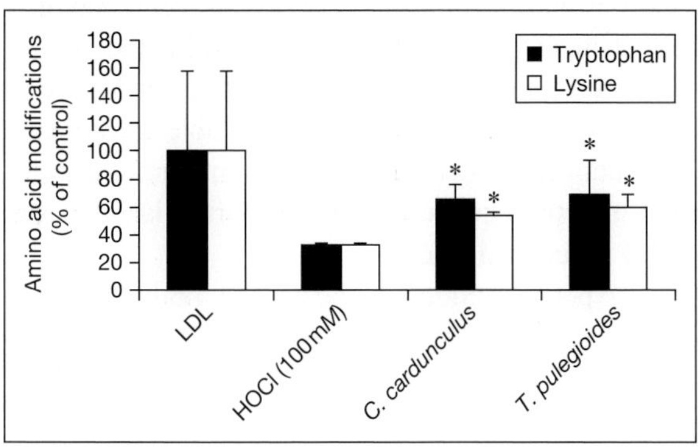

Fig. 2. Tryptophan and lysine loss after incubation of human LDL samples with HOCl, in the presence or absence of local Mediterranean plant food extracts (10^{-5} M, as moles of gallic acid equivalents). Data are means ± SD. *$p < 0.05$ vs. LDL oxidized with 100 μM HOCl. From Schaffer et al. [33].

a loss of tryptophan and lysine. Co-incubation with Mediterranean wild food plant (namely *C. cardunculus* and *T. pulegioides*) extracts strongly protected LDL against the loss of tryptophan residues and partly – at higher concentrations – protected against the loss of lysine [33]. Prevention of the HOCl-induced increase in

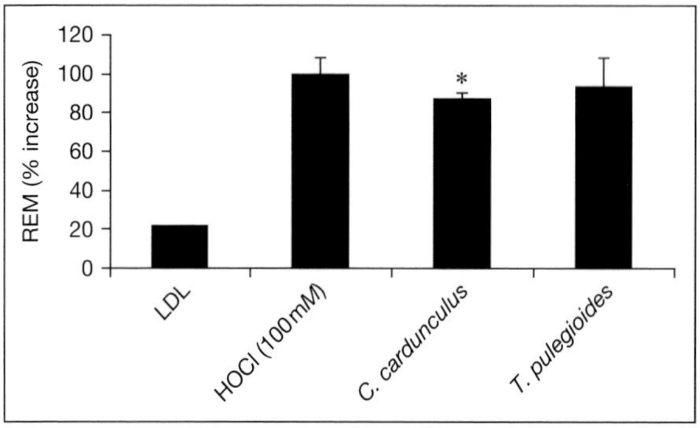

Fig. 3. Effects of local Mediterranean plant food extracts (10^{-5} *M*, as moles of gallic acid equivalents) on HOCl-induced changes in relative electrophoretic mobility (REM) of human LDL. Data are means ± SD. *$p < 0.05$ vs. LDL oxidized with 100 μ*M* HOCl. From Schaffer et al. [33].

REM (fig. 3) was noted for *C. cardunculus* and resveratrol at 10^{-5} *M*, whereas *C. cardunculus*, *T. pulegioides*, lipoic acid, and resveratrol were effective at the concentration of 10^{-6} *M* [33].

In turn, the studies carried out within the Local Food-Nutraceuticals project demonstrate that phenolic-rich extracts of wild plants collected in the Mediterranean area (Lucania – Southern Italy, Crete, and Segura Valley – Southern Spain), namely wild artichoke and wild thyme, are able to increase NO production by stimulated endothelial cells and by a brain homogenate. Also, they are able to scavenge reactive species, HOCl in particular, that are implicated in the development of vascular disease.

This finding adds further experimental evidence to the body of basic research supporting the epidemiological evidence of a lower CHD incidence in the Mediterranean area. In particular, these data suggest that maintenance of correct vasomotion by bioactive components from plants in various Mediterranean regions (Lucania – South Italy, Crete, Segura Valley – Southern Spain) might contribute to its cardioprotective properties. Among the factors that contribute to cardiovascular disease, enhanced oxidative stress leading to modification of LDL or to impaired vascular reactivity does play a role [36], even though the precise nature and the extent of oxidative damage is yet to be fully elucidated. Antioxidant therapy, namely with vitamin E, does not appear to decrease atherosclerosis and cardiovascular mortality [37] and equivocal results were obtained in the treatment of endothelial dysfunction [38]. Conversely, clinical data are supportive of a healthy effect of vitamin C administration [35]. It can be concluded that water-soluble or amphiphilic antioxidants – acting at the lipid-water interface – might effectively scavenge free radicals and/or modulate enzymatic processes relevant to cardiac and vascular functions.

The precise mechanism(s) of action responsible for the enhancement of NO production are still elusive. However, our data exclude any effect on eNOS levels and are in agreement with other data that suggest how water-soluble antioxidants and thiol agents maintain an intracellular reduced environment and preserve cofactors such as BH_4, hence preventing eNOS uncoupling [28] and facilitating its activity [39–42]. Finally, facilitated hydroxylation of eNOS, which is the first step in NO production by this enzyme [43], might also explain the potentiating activities of wild plant extracts, which are rich in *o*-diphenols (we have identified luteolin-7-glicoside, apigenin, rosmaric acid, and ursolic acid in wild artichoke [32]. Ursolic acid is also abundant in nettles and is being studied for its hepatoprotective and antitumorigenic properties [44].

Olive Oil Minor Components

Diets enriched in olive oil result in decreases in overall cardiovascular risk when compared to the STEP II American Heart Association (AHA) diet or a typical Western type diet, while reducing plasma triglyceride levels and not affecting HDL [45]. Monounsaturated fatty acids have been also shown to decrease LDL when substituted for saturated fat [46]. As far as the lipid profile is concerned, olive oil, when consumed on a long-term basis, may also decrease the in vivo plasma oxidized LDL [47] and act favorably on the LDL and HDL subclass distribution [48]. In diabetic subjects, olive oil has been shown to improve the lipid profile [49, 50] and glycemic control [51]. Moreover, in a randomized double-blind study, olive oil was found to reduce blood pressure and the need for antihypertensive drugs in hypertensive individuals, an effect which was corroborated by further descriptive studies [52]. Furthermore, olive oil may have a neutral or modest beneficial effect against various malignancies (e.g. endometrial and ovarian cancers [reviewed in 53].

Research on the impact of olive oil consumption on CHD and mortality has expanded over the last decades, but interest progressively moved from the role of the major component, i.e. oleic acid, to that of the minor constituents, especially those with potent antioxidant properties, i.e. simple and complex phenols, that are present in appreciable concentrations in extra virgin olive oils [54]. In brief, research in this area discovered that phenolic compounds with catecholic structure, namely hydroxytyrosol and oleuropein, particularly abundant in olives and in part released into the oils, are endowed with potent cardioprotective activities [55, 56]. These compounds, due to their physicochemical features, i.e. the relatively high hydrophilic properties, are also released in a significant proportion into the water phase that is generated during the process of oil production (olive mill waste waters) [31]. Recovery of phenols from olive mill waste waters and assessment of their activities has also been part of a research project (FAIR CT 97 3039) sponsored by the EEC. Main outcomes have been reviewed by Visioli et al. [57].

It is noteworthy that hydroxytyrosol and oleuropein are also bioavailable and exert antioxidant (e.g. reduction of urinary isoprostane excretion in animals and humans) and other favorable (antithrombotic and anti-inflammatory) activities in vivo, when administered at doses comparable to those ingested by people in the Mediterranean area [58]. Relevant to nutraceutical formulation, it is worth mentioning that bioavailability of the compounds is greater when ingested as components of extra virgin olive oil rather than when added to phenol-poor oils or yogurt [59], possibly because the water residue in extra virgin olive oil derived from the olive drupe facilitates the suspension and fine dispersion of these amphiphilic compounds in the oils.

Tomato Products and Lycopene

Extensive evidence shows that tomato consumption decreases the incidence of certain types of cancer and other degenerative diseases [60]. Indeed, tomato and its products contain several compounds (vitamin C, folate, polyphenols, etc.) that might modulate radical-mediated oxidative damage that contribute to the initiation and progression of chronic disease processes [60]. Moreover, carotenoids such as lycopene, e.g. β-carotene, phytoene, and phytofluene, are present in considerable amounts both in fresh tomato (~3.5–5 mg/100 g) or its products, e.g. puree (~20 mg/100 g) and paste (~45 mg/100 g).

Lycopene accounts for more than 50% of plasma carotenoids [61] and is also present in other human tissues such as skin, adipose tissue, liver, testes, adrenal gland, and prostate, as well as being abundant in buccal mucosal cells and lymphocytes [60]. High intakes of lycopene are inversely associated with prostate cancer incidence [62] and it is noteworthy that its bioavailability is higher when it is ingested via cooked food, e.g. tomato sauce or pizza, than from raw tomato [63]. Phytoene and phytofluene are also present in human plasma, in concentrations of about 0.1–0.2 μmol/l [64], although only few studies report their concentration in human cells and tissues.

The consumption of tomatoes and tomato products has been shown to increase serum lycopene levels [65, 66]. Most epidemiological studies suggest that high serum lycopene has a protective role against aortic atherosclerosis [67] and myocardial infraction in smokers [68] while low lycopene levels have been associated with acute coronary event or stroke [69] and higher intima-media thickness in men [69, 70]. Moreover, low adipose tissue lycopene levels have been associated with increased risk of myocardial infraction in the multicenter EURAMIC study [71]. Other studies, however, have not confirmed the beneficial effect of lycopene in cardiovascular disease. More specifically, in these studies the association of lycopene and atherosclerosis risk [72], myocardial infraction risk [73] or intima-media thickness [74, 75] was not significant. Intervention studies have shown a lipid-lowering effect of lycopene in doses of 60 mg/day [76] or 20–150 mg/day [65], while its effects on LDL oxidation are controversial [77]. More clinical trials are needed to elucidate the role of lycopene and tomato products on human health.

We carried out intervention trials with tomato and its products, providing healthy volunteers with moderate (7–8 mg/day) amounts of lycopene through whole tomato, puree, and paste. Albeit no increase in plasma antioxidant capacity was recorded, in line with other observations, lower isoprostane excretion and higher resistance of LDL to oxidation was found after supplementation [78, 79]. Interestingly, tomato administration afforded lymphocyte DNA resistance from oxidation, suggesting that the cancer-preventive effects of tomato consumption might indeed be mediated by lycopene and other carotenoids [78, 79]. To further ascertain the role of lycopene in chemoprevention, we undertook an intervention trial with a soft drink (Lyc-o-mato®) added with a lycopene oleoresin that also included phytoene and phytofluene (5.7, 3.7, and 2.7 mg, respectively, in addition to 1 mg of β-carotene and 1.8 mg α-tocopherol/250 ml). We demonstrated that consumption of Lyc-o-mato® increase plasma and lymphocyte carotenoid content and the antioxidant defense system of lymphocytes [80]. Thus, this nutraceutical might constitute an additional or alternative source of bioactive carotenoids.

Concluding Remarks

In conclusion, several food items typical of the Mediterranean area contain compounds for which basic research and intervention trials are demonstrating healthful effects. Formulation of nutraceuticals and functional foods requires supportive evidence that can only come from a research strategy that starts from in vitro experiments, moves on to animal studies and ends with controlled human trials, for which hard endpoints are still actively looked for. Concerning the cardiovascular system, the most relevant issues in nutraceutical research are: (a) the assessments of the equivalence, in terms of bioavailability, of micronutrients administered in different preparations and (b) the correlations between the doses and biomarker levels, i.e. quantitative measurements of parameters that indicate biological (metabolic and functional) processes relevant to cardiovascular physiology. This is an area of growing interest for the potential applications to prevention and treatment of vascular diseases through a 'nutritional' approach that requires, however, adequate experimental designs and updated methodological tools.

Acknowledgments

Research in our laboratory has been mostly supported by EEC grants FAIR CT 97 3039 and QLK-1002-00173. Drs. Shane Thomas and John F. Keaney, Jr., supplied the PAEC. Drs. N. Villeneuve, J.P. Vilaine, and P.M. Vanhoutte provided very helpful advice on NO determination.

References

1 Visioli F, Borsani L, Galli C: Diet and prevention of coronary heart disease: the potential role of phytochemicals. Cardiovasc Res 2000;47:419–425.
2 Iriti M, Faoro F: Plant defense and human nutrition: phenylpropanoids on the menu. Curr Topics Nutr Res 2004;2:47–65.
3 Simopoulos AP, Visioli F (eds): Mediterranean Diets. Basel, Karger, 2000.
4 Simopoulos AP, Salem N Jr: Purslane: a terrestrial source of omega-3 fatty acids. N Engl J Med 1986;315:833.
5 Simopoulos AP, Norman HA, Gillaspy JE, Duke JA: Common purslane: a source of omega-3 fatty acids and antioxidants. J Am Coll Nutr 1992;11:374–382.
6 Simopoulos AP, Norman HA, Gillaspy JE: Purslane in human nutrition and its potential for world agriculture. World Rev Nutr Diet 1995;77:47–74.
7 Zeghichi S, Kallithraka S, Simopoulos AP: Nutritional composition of molokhia (*Corchorus olitorius*) and stamnagathi (*Cichorium spinosum*). World Rev Nutr Diet 2003;91:1–21.
8 Zeghichi S, Kallithraka S, Simopoulos AP, Kypriotakis Z: Nutritional composition of selected wild plants in the diet of Crete. World Rev Nutr Diet 2003;91:22–40.
9 Gerber M: Biofactors in the Mediterranean diet. Clin Chem Lab Med 2003;41:999–1004.
10 Scalbert A, Williamson G: Dietary intake and bioavailability of polyphenols. J Nutr 2000;130: 2073S–2085S.
11 Halliwell B, Whiteman M: Measuring reactive species and oxidative damage in vivo and in cell culture: how should you do it and what do the results mean? Br J Pharmacol 2004;142:231–255.
12 Viegi L, Pieroni A, Guarrera PM, Vangelisti R: A review of plants used in folk veterinary medicine in Italy as basis for a databank. J Ethnopharmacol 2003;89:221–244.
13 Hu FB, Willett WC: Optimal diets for prevention of coronary heart disease. JAMA 2002;288: 2569–2578.
14 Riboli E, Norat T: Epidemiologic evidence of the protective effect of fruit and vegetables on cancer risk. Am J Clin Nutr 2003;78:559S–569S.
15 Trichopoulou A, Costacou T, Bamia C, Trichopoulos D: Adherence to a Mediterranean diet and survival in a Greek population. N Engl J Med 2003;348:2599–2608.
16 Chrysohoou C, Panagiotakos DB, Pitsavos C, Das UN, Stefanadis C: Adherence to the Mediterranean diet attenuates inflammation and coagulation process in healthy adults: the ATTICA Study. J Am Coll Cardiol 2004;44:152–158.
17 De Lorgeril M, Renaud S, Mamelle N, Salen P, Martin JL, Monjaud I, Guidollet J, Touboul P, Delaye J: Mediterranean α-linolenic acid-rich diet in secondary prevention of coronary heart disease. Lancet 1994;343:1454–1459.
18 De Lorgeril M, Salen P, Martin JL, Monjaud I, Boucher P, Mamelle N: Mediterranean dietary pattern in a randomized trial: prolonged survival and possible reduced cancer rate. Arch Intern Med 1998;158:1181–1187.

19 De Lorgeril M, Salen P: Modified Cretan Mediterranean diet in the prevention of coronary heart disease and cancer. World Rev Nutr Diet 2000;87:1–23.
20 Renaud S, de Lorgeril M, Delaye J, Guidollet J, Jacquard F, Mamelle N, Martin JL, Monjaud I, Salen P, Toubol P: Cretan Mediterranean diet for prevention of coronary heart disease. Am J Clin Nutr 1995;61:1360S–1367S.
21 Simopoulos AP, Robinson J (eds): The Omega Diet. New York, HarperCollins, 1999.
22 Parfitt VJ, Rubba P, Bolton C, Marotta G, Hartog M, Mancini M: A comparison of antioxidant status and free radical peroxidation of plasma lipoproteins in healthy young persons from Naples and Bristol. Eur Heart J 1994;15:871–876.
23 Mancini M, Rubba P: The Mediterranean diet in Italy; in Simopoulos AP, Visioli F (eds): Mediterranean diets. World Rev Nutr Diet. Basel, Karger, 2000, pp 114–126.
24 Bellido C, Lopez-Miranda J, Blanco-Colio LM, Perez-Martinez P, Muriana FJ, Martin-Ventura JL, Marin C, Gomez P, Fuentes F, Egido J, Perez-Jimenez F: Butter and walnuts, but not olive oil, elicit postprandial activation of nuclear transcription factor-κB in peripheral blood mononuclear cells from healthy men. Am J Clin Nutr 2004;80:1487–1491.
25 Vita JA, Keaney JF Jr: Endothelial function: a barometer for cardiovascular risk? Circulation 2002;106:640–642.
26 Assanelli D, Bonanome A, Pezzini A, Albertini F, Maccalli P, Grassi M, Archetti S, Negrini R, Visioli F: Folic acid and vitamin E supplementation effects on homocysteinemia, endothelial function and plasma antioxidant capacity in young myocardial-infarction patients. Pharmacol Res 2004;49:79–84.
27 Thomas SR, Chen K, Keaney JF Jr: Oxidative stress and endothelial nitric oxide bioactivity. Antioxid Redox Signal 2003;5:181–194.
28 Cai H, Harrison DG: Endothelial dysfunction in cardiovascular diseases: the role of oxidant stress. Circ Res 2000;87:840–844.
29 Schulz E, Anter E, Keaney JF Jr: Oxidative stress, antioxidants, and endothelial function. Curr Med Chem 2004;11:1093–1104.
30 Carr A, Frei B: The role of natural antioxidants in preserving the biological activity of endothelium-derived nitric oxide. Free Radic Biol Med 2000;28:1806–1814.
31 Visioli F, Vinceri FF, Galli C: 'Waste waters' from olive oil production are rich in natural antioxidants. Experientia 1995;51:32–34.
32 Grande S, Bogani P, de Saizieu A, Schueler G, Galli C, Visioli F: Vasomodulating potential of Mediterranean wild plant extracts. J Agric Food Chem 2004;52:5021–5026.
33 Schaffer S, Eckert GP, Muller WE, Llorach R, Rivera D, Grande S, Galli C, Visioli F: Hypochlorous acid scavenging properties of local Mediterranean plant foods. Lipids 2004;39:1239–1247.
34 Rossoni G, Grande S, Galli C, Visioli F: Wild artichoke prevents of age-associated loss of vasomotor function. J Agric Food Chem 2005;53:10291–10296.
35 Carr AC, Tijerina T, Frei B: Vitamin C protects against and reverses specific hypochlorous acid- and chloramine-dependent modifications of low-density lipoprotein. Biochem J 2000;346:491–499.
36 Diaz MN, Frei B, Vita JA, Keaney JF Jr: Antioxidants and atherosclerotic heart disease. N Engl J Med 1997;337:408–416.
37 Kritharides L, Stocker R: The use of antioxidant supplements in coronary heart disease. Atherosclerosis 2002;164:211–219.
38 Visioli F: Effects of vitamin E on the endothelium: equivocal? Alpha-tocopherol and endothelial dysfunction. Cardiovasc Res 2001;51:198–201.
39 Heller R, Munscher-Paulig F, Grabner R, Till U: *L*-Ascorbic acid potentiates nitric oxide synthesis in endothelial cells. J Biol Chem 1999;274:8254–8260.
40 Heller R, Unbehaun A, Schellenberg B, Mayer B, Werner-Felmayer G, Werner ER: *L*-Ascorbic acid potentiates endothelial nitric oxide synthesis via a chemical stabilization of tetrahydrobiopterin. J Biol Chem 2001;276:40–47.
41 Huang A, Vita JA, Venema RC, Keaney JF Jr: Ascorbic acid enhances endothelial nitric-oxide synthase activity by increasing intracellular tetrahydrobiopterin. J Biol Chem 2000;275:17399–17406.
42 Huang A, Xiao H, Samii JM, Vita JA, Keaney JF Jr: Contrasting effects of thiol-modulating agents on endothelial NO bioactivity. Am J Physiol 2001;281:C719–C725.

43 Moncada S, Palmer RM, Higgs EA: Nitric oxide: physiology, pathophysiology, and pharmacology. Pharmacol Rev 1991;43:109–142.
44 Horvathova K, Vachalkova A, Novotny L: Flavonoids as chemoprotective agents in civilization diseases. Neoplasma 2001;48:435–441.
45 Kris-Etherton PM, Pearson TA, Wan Y, Hargrove RL, Moriarty K, Fishell V, Etherton TD: High-monounsaturated fatty acid diets lower both plasma cholesterol and triacylglycerol concentrations. Am J Clin Nutr 1999;70:1009–1015.
46 Mensink RP, Zock PL, Kester AD, Katan MB: Effects of dietary fatty acids and carbohydrates on the ratio of serum total to HDL cholesterol and on serum lipids and apolipoproteins: a meta-analysis of 60 controlled trials. Am J Clin Nutr 2003;77:1146–1155.
47 Marrugat J, Covas MI, Fito M, Schroder H, Miro-Casas E, Gimeno E, Lopez-Sabater MC, de la Torre R, Farre M: Effects of differing phenolic content in dietary olive oils on lipids and LDL oxidation – a randomized controlled trial. Eur J Nutr 2004;43:140–147.
48 Zambon A, Sartore G, Passera D, Francini-Pesenti F, Bassi A, Basso C, Zambon S, Manzato E, Crepaldi G: Effects of hypocaloric dietary treatment enriched in oleic acid on LDL and HDL subclass distribution in mildly obese women. J Intern Med 1999;246:191–201.
49 Madigan C, Ryan M, Owens D, Collins P, Tomkin GH: Dietary unsaturated fatty acids in type 2 diabetes: higher levels of postprandial lipoprotein on a linoleic acid-rich sunflower oil diet compared with an oleic acid-rich olive oil diet. Diabetes Care 2000;23:1472–1477.
50 Thomsen C, Storm H, Holst JJ, Hermansen K: Differential effects of saturated and monounsaturated fats on postprandial lipemia and glucagon-like peptide 1 responses in patients with type 2 diabetes. Am J Clin Nutr 2003;77:605–611.
51 Garg A: High-monounsaturated-fat diets for patients with diabetes mellitus: a meta-analysis. Am J Clin Nutr 1998;67:577S–582S.
52 Psaltopoulou T, Naska A, Orfanos P, Trichopoulos D, Mountokalakis T, Trichopoulou A: Olive oil, the Mediterranean diet, and arterial blood pressure: the Greek European Prospective Investigation into Cancer and Nutrition (EPIC) Study. Am J Clin Nutr 2004;80:1012–1018.
53 Trichopoulou A, Georgiou E, Bassiakos Y, Lipworth L, Lagiou P, Proukakis C, Trichopoulos D: Energy intake and monounsaturated fat in relation to bone mineral density among women and men in Greece. Prev Med 1997;26:395–400.
54 Montedoro GF, Servili M, Baldioli M, Miniati E: Simple and hydrolyzable phenolic compounds in virgin olive oil. 1. Their extraction, separation, and quantitative and semiquantitative evaluation by HPLC. J Agric Food Chem 1992;40:1571–1576.
55 Visioli F, Galli C: Biological properties of olive oil phytochemicals. Crit Rev Food Sci Nutr 2002;42:209–221.
56 Perez-Jimenez F: International conference on the healthy effect of virgin olive oil. Eur J Clin Invest 2005;35:421–424.
57 Visioli F, Poli A, Gall C: Antioxidant and other biological activities of phenols from olives and olive oil. Med Res Rev 2002;22:65–75.
58 Visioli F, Bogani P, Galli C: Healthful properties of olive oil minor components; in Boskou D (ed): Olive Oil Chemistry and Technology. Champaign, American Oil Chemists Society Press, 2006.
59 Visioli F, Galli C, Grande S, Colonnelli K, Patelli C, Galli G, Caruso D: Hydroxytyrosol excretion differs between rats and humans and depends on the vehicle of administration. J Nutr 2003;133:2612–2615.
60 Rao AV, Rao LG: Lycopene and human health. Curr Topics Nutr Res 2004;2:127–136.
61 Johnson EJ: The role of carotenoids in human health. Nutr Clin Care 2002;5:56–65.
62 Giovannucci E, Rimm EB, Liu Y, Stampfer MJ, Willett WC: A prospective study of tomato products, lycopene, and prostate cancer risk. J Natl Cancer Inst 2002;94:391–398.
63 Gartner C, Stahl W, Sies H: Lycopene is more bioavailable from tomato paste than from fresh tomatoes. Am J Clin Nutr 1997;66:116–122.
64 Hoppe PP, Kramer K, van den BH, Steenge G, van Vliet T: Synthetic and tomato-based lycopene have identical bioavailability in humans. Eur J Nutr 2003;42:272–278.
65 Rao AV, Agarwal S: Bioavailability and in vivo antioxidant properties of lycopene from tomato products and their possible role in the prevention of cancer. Nutr Cancer 1998;31:199–203.

66 Hadley CW, Clinton SK, Schwartz SJ: The consumption of processed tomato products enhances plasma lycopene concentrations in association with a reduced lipoprotein sensitivity to oxidative damage. J Nutr 2003;133:727–732.
67 Klipstein-Grobusch K, Launer LJ, Geleijnse JM, Boeing H, Hofman A, Witteman JC: Serum carotenoids and atherosclerosis. The Rotterdam Study. Atherosclerosis 2000;148:49–56.
68 Street DA, Comstock GW, Salkeld RM, Schuep W, Klag MJ: Serum antioxidants and myocardial infarction. Are low levels of carotenoids and α-tocopherol risk factors for myocardial infarction? Circulation 1994;90:1154–1161.
69 Rissanen T, Voutilainen S, Nyyssonen K, Salonen JT: Lycopene, atherosclerosis, and coronary heart disease. Exp Biol Med (Maywood) 2002;227:900–907.
70 Rissanen T, Voutilainen S, Nyyssonen K, Salonen R, Salonen JT: Low plasma lycopene concentration is associated with increased intima-media thickness of the carotid artery wall. Arterioscler Thromb Vasc Biol 2000;20:2677–2681.
71 Kohlmeier L, Kark JD, Gomez-Gracia E, Martin BC, Steck SE, Kardinaal AF, Ringstad J, Thamm M, Masaev V, Riemersma R, Martin-Moreno JM, Huttunen JK, Kok FJ: Lycopene and myocardial infarction risk in the EURAMIC Study. Am J Epidemiol 1997;146:618–626.
72 D'Odorico A, Martines D, Kiechl S, Egger G, Oberhollenzer F, Bonvicini P, Sturniolo GC, Naccarato R, Willeit J: High plasma levels of α- and β-carotene are associated with a lower risk of atherosclerosis: results from the Bruneck study. Atherosclerosis 2000;153:231–239.
73 Hak AE, Stampfer MJ, Campos H, Sesso HD, Gaziano JM, Willett W, Ma J: Plasma carotenoids and tocopherols and risk of myocardial infarction in a low-risk population of US male physicians. Circulation 2003;108:802–807.
74 Iribarren C, Folsom AR, Jacobs DR Jr, Gross MD, Belcher JD, Eckfeldt JH: Association of serum vitamin levels, LDL susceptibility to oxidation, and autoantibodies against MDA-LDL with carotid atherosclerosis. A case-control study. The ARIC Study Investigators. Atherosclerosis Risk in Communities. Arterioscler Thromb Vasc Biol 1997;17:1171–1177.
75 Bonithon-Kopp C, Coudray C, Berr C, Touboul PJ, Feve JM, Favier A, Ducimetiere P: Combined effects of lipid peroxidation and antioxidant status on carotid atherosclerosis in a population aged 59–71 years: the EVA Study. Etude sur le Vieillisement Arteriel. Am J Clin Nutr 1997;65:121–127.
76 Fuhrman B, Elis A, Aviram M: Hypocholesterolemic effect of lycopene and β-carotene is related to suppression of cholesterol synthesis and augmentation of LDL receptor activity in macrophages. Biochem Biophys Res Commun 1997;233:658–662.
77 Hininger IA, Meyer-Wenger A, Moser U, Wright A, Southon S, Thurnham D, Chopra M, van den BH, Olmedilla B, Favier AE, Roussel AM: No significant effects of lutein, lycopene or β-carotene supplementation on biological markers of oxidative stress and LDL oxidizability in healthy adult subjects. J Am Coll Nutr 2001;20:232–238.
78 Visioli F, Riso P, Grande S, Galli C, Porrini M: Protective activity of tomato products on in vivo markers of lipid oxidation. Eur J Nutr 2003;42:201–206.
79 Riso P, Visioli F, Erba D, Testolin G, Porrini M: Lycopene and vitamin C concentrations increase in plasma and lymphocytes after tomato intake. Effects on cellular antioxidant protection. Eur J Clin Nutr 2004;58:1350–1358.
80 Porrini M, Riso P, Brusamolino A, Berti C, Guarnieri S, Visioli F: Daily intake of a formulated tomato drink affects carotenoid plasma and lymphocyte concentrations and improves cellular antioxidant protection. Br J Nutr 2005;93:93–99.

Francesco Visioli
Department of Pharmacological Sciences
Via Balzaretti 9, IT–20133 Milan (Italy)
Tel. +39 02 5031 8280, Fax +39 02 700426 106, E-Mail francesco.visioli@unimi.it

Heinrich M, Müller WE, Galli C (eds): Local Mediterranean Food Plants and Nutraceuticals. Forum Nutr. Basel, Karger, 2006, vol 59, pp 130–153

A Vegetable to Meat Consumption Ratio as a Relevant Factor Determining Cancer Preventive Diet

The Mediterranean versus Other European Countries

Maria Kapiszewska

Department of General Biochemistry, Faculty of Biotechnology, Jagiellonian University, Krakow, Poland

Abstract

The observed growth of cancer incidence in certain regions has been usually linked to frequent consumption of 'unhealthy' food. Such food often contains genotoxic substances as heterocyclic aromatic amines (HAAs) and/or polycyclic aromatic hydrocarbones (PAHs), occurring during food preparation, which induce DNA damage in cells. These substances are mainly formed during frying or grilling of meat and they can be removed from the body in a two-stage metabolic process of detoxification (phase 1 and phase 2). If they are not excreted, they form DNA adducts. The effectiveness of detoxification depends on the activity of enzymes encoded by polymorphic genes. A diet containing plenty of fruits and vegetables, due to the presence of biologically active polyphenols, can modulate activity of detoxifying enzymes. Such a diet can decrease the extent of DNA adducts, breaks and oxidative damage, supporting the body's enzymatic system in sufficient removal of DNA damage. The antioxidant vitamins' content in such a diet also enhances the DNA protection by increasing the scavenging of radical oxidative species that occurs during metabolic reactions. The lack of balance between the amount of 'unhealthy' and 'healthy' food leads to the accumulation of unrepaired damage, initiating DNA instability and inducing cancer development. Such damage is often used as a biomarker of cancer risk in epidemiological studies. Moreover, in in vitro studies, the amount of the DNA damage is used as indicator of the protective ability of vitamins, plant extracts and/or individual flavonoids. The incidences of certain dietary-related cancers in European Mediterranean countries is lower than in Central and Northern European countries; there is simultaneously variation in the habitual diet in these regions. This suggests that some features of routine nutrition in the Mediterranean countries may be responsible for this preventing effect. However, inconsistency in the epidemiological data, associating the meat and fruit and vegetable intake with cancer risk, suggests that another strategy for evaluation dietary influence on cancer risk should be undertaken. This article argues that it is not the consumption of a single food product or an individual component of diet, but rather a proper ratio of vegetable to meat consumption that is responsible for cancer prevention. This hypothesis is tested comparing the association between certain dietary-related cancer incidences (colon & rectum, breast and prostate cancer), registered in 2002, with the ratio between consumption of these two groups of food products in the Mediterranean region and in Central and Northern European region over the last three decades. The results clearly showed that both the ratio between vegetables and meat consumption as well as the ratio between the amount of energy from vegetables and from animal products can be used successfully to evaluate the dietary pattern related to cancer risk.

Differences in dietary patterns between the countries of Mediterranean Europe and other European countries have been analyzed extensively [1–9]. Studies have demonstrated that diets in most countries of Northern Europe are high in animal fats (i.e. high in saturated fatty acids), high in simple sugars and low in fibers, mainly because of relatively low consumption of vegetables. In turn, in the countries of Southern Europe, the traditional diet is rich in olive oil, fruits, and vegetables, certain legumes in particular; moreover, a high consumption of food of plant origin there provides an abundant amount of fibers, polyphenols and other antioxidants. In addition, the typical diet in Mediterranean Europe has always included more cereals, pulses and fish but less saturated fatty acids than has the Northern Europe diet. The documented benefits of the Mediterranean dietary patterns include a lower risk of cancer and coronary heart disease, most probably thanks to modulation of certain metabolic pathways, including processes repairing DNA damage [3, 10–19]. It appears that the detoxifying properties of the bioactive compounds in fruits and vegetables can overcome the carcinogenous effect of genotoxic substances present in well-done meat and meat products.

The quantitative differences in vegetable and fruit intake are not the only factors determining the extent of the protective effect. The consumption of staples, roots and wild food plants may have a strong protective effect, although they are consumed on a less regular basis. That ability may result from their different phytochemical profile, as compared to the fruit and vegetable one. This issue was addressed in the EU's project 'Local Food and Nutraceuticals'. Over 100 plants, traditionally used as dietary by-products in rural communities of Southern Italy, Greece and Southern Spain, were investigated for their antioxidant, antidiabetic, memory-mediating and DNA-protective activity [9].

Only in some epidemiological studies was high meat and animal fat intake associated with risk of colorectal, breast and prostate cancer, whereas in other studies such a relationship was not observed [20–25]. However, in the last few decades, the consumption of meat and meat products has increased considerably in Mediterranean countries but the cancer risk has remained low [26]. This reality called into question the assertion that meat is the main dietary factor enhancing cancer risk. Therefore, it seems important to address certain issues, for example, to what extent consumption of the proper proportion of different groups of food products is more important than the amount of consumed food of any one type, or to what extent cancer development depends on the period, in terms of lifespan, in which intake of certain 'unhealthy' food crosses a certain level.

In this paper these problems are analyzed by exploring the relationship between dietary patterns and cancer rates at the population level. An advantage of this kind of analysis is that it can overcome effects of intra-population ranges in food intake and dietary constituencies; such ranges are sometimes broader than inter-population ones. What is even more important is the stability of the national diet over time, as compared to the variations in diet among individuals. At the same time, it is necessary to keep in mind that data from the population studies represent the quantity of food available for consumption and may overstate actual food intake. An additional aspect of the issue of the putative effects of plant foods, which has been addressed in this chapter, concerns their mechanisms, on the basis of experimental results obtained in the context of the mentioned already the EU's project 'Local Food and Nutraceuticals'.

Methodological Issues

The analysis of relations between consumption of two food groups: (1) meat and meat products, and (2) vegetables and their products, and, on the other hand, cancer risk became the aim of this study. It is assumed that a high consumption of meat and meat products enhances the risk of several cancers, whereas diet high in vegetables has a protective effect. Based on this, the dietary recommendations were established. But the general dietary patterns, beneficial for health, are not precise enough. Assuming that genotoxic substances in the processed meat can be neutralized by consumption of a certain amount of vegetables, the lower cancer incidences should occur in countries where such beneficial meal patterns exist. Do differences exist in the vegetable and meat consumption ratio between Mediterranean and other European countries?

To answer this question the consumption of meat, fruits, vegetables and their products, as well as the percentage of fat and protein energy from certain food products, was correlated with the colon & rectum, breast and prostate cancer incidences, cancers strongly attributable to diet. Cancer incidences for an Age-Standardized Rate (ASR(W), World standard) were taken from 'the GLOBOCAN 2002' database electronically available, which was built up using the data from the Descriptive Epidemiology Group of IARC (www-dep.iarc.fr/globocan/database.htm). GLOBOCAN 2002 presents estimates for the year 2002 but the actual disease rates are 2–5 years earlier. Only very recent cancer data were taken into account due to the fact that the quality of diagnosis and consistent recording of cancer incidence are much better now than they were in the past. Moreover, the evaluation of diseases is much more uniform among countries now. The advantage of using cancer-incidence data rather than often used mortality statistics is that the former are less influenced by the efficiency of therapeutic care. On the other hand, a weakness of cancer-incidence data is that the intensity of screening varies greatly among countries. Therefore, cancer incidence might be overestimated in some areas and underestimated in others. Nevertheless, the data were generated from large populations and were consequently subject to only small random errors.

The food data were obtained from the UN Food and Agriculture Organization's (FAO) (www.fao.org/faostat) for the periods 1969–71, 1975–76, 1980–81, 1990–95 and 2000–2002, as cancer usually develops for years before it is clinically recognized and registered.

Eighteen countries (named in the text as ALL) were chosen for the analysis: seven from the European Mediterranean region (MED) (Greece, Italy, Spain, Portugal, France, Cyprus, and Turkey) and eleven from Central and Northern Europe (OTHERS) (Austria, Denmark, Finland, Germany, Iceland, Ireland, the Netherlands, Norway, Sweden, Switzerland, and the United Kingdom). The two groups of countries were chosen on the basis of the assumed differences in diet patterns and cancer incidences between regions.

Descriptive statistics in the text and tables are given as the mean ± SD (if not indicated otherwise). Statistical analysis was conducted using the Statistica Version 6. The Pearson correlation was applied for correlation analysis. The analysis of variance (ANOVA) was applied to determine the relationship between the country groups regarding the consumption of food products and cancer incidences.

Meat and Meat Product Consumption and Cancer Risk

The consumption of meat and meat products has continuously increased over approximately the last 30 years in each of the 18 countries studied. The mean value of 155 ± 49 g/person/day in the 1969–71 period had increased to 233 ± 63 g/person/day by 2002 in ALL.

In MED, where meat consumption was lowest in 1969–71 among all the countries analyzed, meat consumption increased by 75%, whereas in OTHERS meat consumption increased by only 40%. The meat and meat product consumption was not significantly different between the two groups of countries at any time over the three decades analyzed; the mean values were 192 ± 80 g/person/day for MED and 207 ± 46 g/person/day for OTHERS. The meat and meat product intake in Turkey remains the lowest in Europe (the mean value is now 52 g/person/day). Spain, where the level of meat consumption was one of the lowest in 1969 (122 g/day/person), has become the biggest consumer (322 g/day/person). This analysis illustrates the significant changes in the consumption of meat and meat products that have occurred in the last three decades and underline the differences in the dietary pattern between MED and OTHERS (table 1). In spite of these changes, the consumption of meat and meat products was significantly positively correlated with the incidence of colon & rectum and breast cancer in ALL, as well as in MED, but not in OTHERS. The meaningful difference was found when the timing of food consumption was taken into consideration. The breast cancer incidences were significantly correlated only with consumption of meat in the early years (two and three decades before breast cancer was diagnosed) (table 2).

For further analysis, the countries were divided into two groups based on their consumption of meat and meat products (below and above medium value). The median values increased from 189 to 277 g/day/person over three decades. Only Turkey and Greece remained in the 'lower consumption' group in all analyzed periods. The group with 'lower consumption' of meat and meat products in the 1969–71 and 1980–81 periods had a lower number of colon & rectum ($p = 0.04$ and 0.008, respectively) and breast cancer incidences ($p = 0.03$ and 0.02, respectively). The population in countries that consumed meat and meat products above the mean value, which was 225 g/person/day, in the above-mentioned periods had a higher risk of both cancers than did countries in which less meat was consumed. No correlation was found in prostate cancer risk in ALL but a positive correlation occurred in MED.

Meat Preparation Consequences and Genetic Susceptibility

In addition to the quantity of meat consumed by individuals, the quality determined by the amount of fat and proteins is also important for cancer risk. The cooking method (fried, barbequed, flame-broiled), kind of meat consumed, cooking temperature, and the duration of the process determine the amount and type of carcinogenous substances [27]. Also, metabolism of fat and oils is the additional source of DNA damage through the formation of reactive oxygen species [28–31]. Most substances formed during the cooking of animal proteins are mutagenic and carcinogenic [32]. It is possible that a high fat diet in combination with a heterocyclic amine (HCA) carcinogen derived from cooked meat may enhance the incidence and severity of cancer as was shown for mammary gland cancer in rats [33]. An excellent review of the history and importance of aromatic amine and HCA in public health was recently presented by Weisburger [34]. The

Table 1. Mean values (g/person/day) of consumption of certain food groups: meat (M), vegetables (V) and fruits (F) (and their products), the percentage of energy from animal and vegetable products, and the ratio between vegetables & products and meat & products consumption in ALL, MED and OTHERS countries. The FAO data from certain years and GLOBOCAN data for cancer incidences in 2002

Countries	OTHERS mean ± SD	MED mean ± SD	ALL mean ± SD	MED vs. OTHERS, p
ASR(W) breast cancer incidences	82.4 ± 6.60	59.1 ± 22.0	73.4 ± 18.2	0.00405
ASR(W) prostate cancer incidences	68.3 ± 16.3	37.4 ± 16.5	56.3 ± 22.2	0.00124
ASR(W) colon % rectum cancer incidences	67.8 ± 9.50	49.2 ± 18.2	60.6 ± 16.1	0.01142
% of energy from animal products (%A)	35.5 ± 3.10	25.5 ± 5.40	31.6 ± 6.4	0.00013
% of energy from vegetable products (%V)	64.5 ± 3.10	74.5 ± 5.40	68.4 ± 6.4	0.00013
%V/%A	1.83 ± 0.20	3.04 ± 0.70	2.30 ± 0.8	0.00004
M 1969–71	171 ± 36	130 ± 60	155 ± 49	0.03268
M 1979–81	203 ± 39	167 ± 73	189 ± 56	0.22794
M 1990–92	215 ± 48	206 ± 74	211 ± 58	0.45213
M 1995–97	217 ± 46	219 ± 78	218 ± 58	0.75643
M 2000–02	230 ± 46	238 ± 86	233 ± 63	0.75547
V 1969–71	144 ± 63	428 ± 82	254 ± 158	0.00000
V 1979–81	177 ± 61	439 ± 150	279 ± 165	0.00008
V 1990–92	193 ± 45	499 ± 131	312 ± 175	0.00000
V 1995–97	205 ± 37	501 ± 132	320 ± 170	0.00000
V 2000–02	223 ± 42	508 ± 123	334 ± 164	0.00000
Cereals & prod. excl. beer 1969–71	269 ± 44	404 ± 110	322 ± 100	0.00207
Cereals & prod. excl. beer 79/80	259 ± 37	384 ± 128	308 ± 102	0.00698
Cereals & prod. excl. beer 90/91	270 ± 43	386 ± 125	315 ± 100	0.01107
Cereals & prod. excl. beer 95/96	275 ± 43	381 ± 112	316 ± 91	0.01118
Cereals & prod. excl. beer 2000–02	297 ± 45	385 ± 115	331 ± 88	0.03379
Pulses & products 1969–71	4.00	18.1 ± 6.3	9.5 ± 8.3	0.00001
Pulses & products 79/80	4.30	14.1 ± 5.3	8.1 ± 6.1	0.00004
Pulses & products 90/91	5.20	17.6 ± 11	10 ± 9.3	0.00257
Pulses & products 95/96	5.10	15.9 ± 9	9.3 ± 8.2	0.00277
Pulses & products 2000–02	4.80	15.1 ± 8.9	8.8 ± 8	0.00361
V 1969–71/M	0.84	4.20 ± 2.8	2.15 ± 2.4	0.00107
V 1979–81/M	0.87	3.80 ± 3.8	2.01 ± 2.7	0.02055
V 1990–92/M	0.97	3.34 ± 3.1	1.89 ± 2.2	0.02084
V 1995–97/M	0.96	3.39 ± 3.7	1.91 ± 2.5	0.04033
V 2000–02/M	0.99	3.28 ± 3.8	1.88 ± 2.5	0.05749

heterocyclic aromatic amines (HAAs), such as 2-amino-3-methylimidazo [4,5-f] quinoline [35] and phenylimidazopyridine, identified in cooked meat, and polycyclic aromatic hydrocarbones (PAHs), are genotoxic [36, 37]. In addition, high-meat diets in humans increase the level of nitrosatable material entering the colon, so fecal *N*-nitroso compounds increase in a dose-responsive manner [38]. All of

Table 2. The statistically significant, positive and negative Pearson correlation (+* and −* respectively) between, on the one hand: (1) meat & products (g/day/person); (2) vegetables & products; (3) energy from meat fat (%EF); (4) energy from meat protein (%EP); (5) energy from milk & products' protein (%EP); (6) energy from dietary animal products (APDE); (7) cereals & products (g/day/person); (8) pulses & products; (9) the ratio between vegetables & products to meat & products, and, on the other hand: colon & rectum, breast and prostate cancer incidences, ASR(W), registered in 2002

	ALL			MED			OTHERS		
	colon & rectum	breast	prostate	colon & rectum	breast	prostate	colon & rectum	breast	prostate
Meat & products	+*	+* 69/81	NO	+* 79/02	+*	+*	NO	NO	+* 90/02
%EF from meat & products	NO	+*	+*	+* 79/02	+*	+*	NO	NO	NO
%EP from meat & products	+*	+*	+*	+*	+*	+*	NO	NO	NO
%E milk & products	+*	+*	+*	NO	+*	NO	NO	NO	NO
%APDE	+*	+*	+*	NO	+* 79/02	+* 90/02	NO	NO	NO
Vegetables & products	−*	−*	−*	−* 00/02	NO	NO	NO	NO	NO
Cereals & products	−*	−*	−*	NO	NO	−*	NO	−* 69/90	NO
Pulses & products	−*	−*	−*	−*	−*	−* 79/02	NO	−* 69/81	NO
V/M	−*	−*	−*	−*	−*	−*	NO	NO	NO

For the above calculations, mean values of all dietary factors from the period 1969–2002 were used (to be exact from years: 1969–71, 1979–81, 1990–91, 1995–97, and 2000–2002). Lack of statistically significant association is labeled NO. Information about particular periods is provided when correlation is statistically significant only in this period. The data were taken from the FAO, while cancer incidences were taken from GLOBOCAN 2002 (IARC).

them reveal carcinogenic properties if the balance between the ability to metabolize them to electrophilic, reactive metabolites in phase 1 and the rate of formation of hydrophilic conjugation products by phase 2 enzymes is disturbed [39]. Most of those enzymes belong to a large family of isozymes involved in the detoxification of many electrophilic substrates by their conjugation. Several of the genotypes of these isozymes seem to be associated with different levels of expression of enzyme activity. Among them is the arylamide *N*-acetyltransferase

(NAT), which requires acety-coenzymes A as a cofactor for its activity. NAT is widely expressed throughout the body [40]. Several studies showed that NAT1 and NAT2 acetylation polymorphisms played a certain role in the cancer risk in human populations [41, 42]. The activity of these NAT isoenzymes is determined primarily by multiple genetically determined variants [43, 44]. The enzymes differ significantly in their intrinsic stabilities and acceptor substrate selectivity and are also involved in meat-derivate carcinogen detoxification [45]. Variants of these genes are described as fast and slow acetylators. Several subtypes of slow acetylators have been identified [46, 47]. Slow NAT2 acetylators carry two germline copies of any of several mutant NAT2 alleles. The heterozygous individuals with NAT2 polymorphisms have a rapid/intermediate acetylator phenotype while those without polymorphism have a slow phenotype.

In first step, the HAAs present in fried meat are metabolically activated through *N*-hydroxylation by cytochrome P_{450}1A2 and further O-acetylated (esterified) by the NAT enzymes to reactive forms, which produce DNA adducts [48]. The amino groups and C-8 atoms of purine nucleosides in DNA seem to be the most prevalent target for transfer of the HCA residue forming DNA adducts in various tissues [49–51]. The accumulation of unrepaired or misrepaired DNA damage may lead to development of cancer. The degree of cancer risk related to variation in functional polymorphism of the NAT gene may also depend on the proportion of different substances formed in cooked meat. The NAT2 slow acetylators are at increased risk for substances for which *N*-acetylation is a detoxification step. In turn, rapid acetylators would be expected to be at greater risk for substances for which O-acetylation is an activation step [41, 42]. Moreover, additional amounts of HAAs (gained through cigarette smoking, for example) can enhance the risk of colorectal cancer among those who prefer to consume well-done meat, especially individuals who have a rapid phenotype for both NAT2 and CYP1A2 [52–54]. However, it is not clear whether meat, red meat, HCAs, or PAHs are associated with the risk of rectal cancer.

Polymorphic enzymes, such as sulfotransferase, NAD(P)H:quinone oxidoreductase or catechol-*O*-methyltransferase, are also involved in the process of detoxification [14]. The glutathione S-transferase (GST) enzyme and its variants are responsible for detoxification of another potential carcinogen formed during meat cooking: the HAAs. For example, the GST(M1) variant with no activity is associated with altered risk of cancer [55, 56]. The combination of polymorphic variants that encode a low activity allele in individuals can significantly influence responsiveness to the nutritional environment, resulting in insufficient detoxification, thus enhancing the genotoxicity of food xenobiotics.

Although numerous epidemiological studies have tried to determine the involvement of gene polymorphisms in the susceptibility of cells to certain dietary compounds, direct and straightforward evidence of the relation between

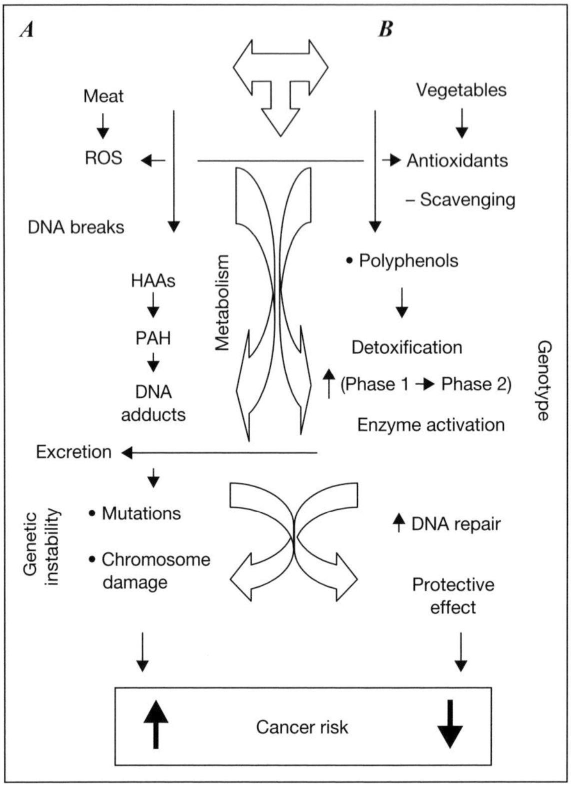

Fig. 1. Meat-vegetables cancer relations: (***A***) consumption of fried/grilled meat inducing DNA damage and enhancing cancer risk due to presence of the polycyclic aromatic hydrocarbons (PAHs), heterocyclic aromatic amines (HAAs), reactive oxygen species (ROS); (***B***) consumption of vegetables increasing the scavenging of ROS and modulating activity of detoxifying enzymes of phase 1 and phase 2 due to antioxidant vitamins and polyphenols and resulting in lowering cancer risk.

these factors is still lacking. This is probably the case because there are potentially millions of base pair differences between individuals and many of them can affect the way in which individuals respond to their nutritional environment. The most accurate data seem to come from subjects who share great sensitivity, in general individuals from the same family [57, 58] or from ethnic groups in which the frequency of certain polymorphism is greater than in other groups.

Summarizing: the genetic background, amount of meat intake and differences in the amount and type of genotoxic substances generated during meat preparation due to meat quality (the percentage of fat and proteins) can determine the susceptibility of individuals to carcinogens (fig. 1).

Meat Constituencies and Cancer Risk

A growing number of studies arising from around the world purport to show a link between fat, saturated fat and/or cholesterol or food groups rich in these constituencies (i.e. dairy products and/or red meat consumption) and several types of cancer [59, 60]. Combined estimates of risk for total and saturated fat intake, and for meat intake, all indicate an association between higher intake and an increased risk of breast cancer based on a meta-analysis [61].

In the research described in this paper, the correlation analysis was used to test the relationship between the meat fat (%EF) and protein (%EP) intake (as a contribution to total dietary fat and protein consumption) and cancer incidences. The %EF was not statistically different between MED and OTHERS, as well as the %EP, except for the 1969–71 period when it was significantly higher in OTHERS. The colon & rectum cancer incidences were not associated with %EF in ALL and in OTHERS, but were positively associated in MED. In turn, colon & rectum, breast and prostate cancer incidences and %EP were positively correlated in ALL and MED (table 2). Consequently, it can be assumed that the consumption of HAAs formed via the Maillard reaction during frying or grilling of protein-rich food of animal origin can be more carcinogenic than PAHs deposited in food from fat dripping onto the burning coals during this type of cooking [62, 63].

A high correlation between national per capita disappearance of fat and national rates of colon cancer led to the hypothesis that consumption of fat, especially from animal sources, increases risk for colon cancer. It was suggested by Giovannucci and Goldin [64] that total lipid content or other factors, such as HCAs formed during cooking, rather than the red meat consumption, is associated with colon carcinogenesis. That appears to be in full agreement with results presented in this paper.

The differences in meat preparation preferences between the two studied regions can both have an impact on cancer risk. For example, in ALL, the prostate cancer incidence seems to depend more on %EF and %EP than on the amount of meat consumed, whereas breast cancer appears to depend not only on meat quantity but also on its quality (table 2).

These results appear to confirm the conclusion drawn from critical evaluation of 33 published case-control and cohort studies which showed that the specific types of fatty foods (e.g. milk or meat) are associated with prostate cancer [65].

In the research presented here, the regression analysis showed increased colon & rectum cancer incidences with an increased percentage of energy consumed with animal products (%APDE) in ALL in each study period (p varied from 0.04 in 1969–71 to 0.01 in 1995–96). In addition, the lack of such a dependence for %EF and significant dependence for %EP from meat and meat products, suggest that proteins, and eventually changes associated with the product preparation, may play a considerable role here.

Moreover, the percentage of proteins from milk & products (%EP milk & products) can have a significant positive impact on breast cancer risk in ALL, as well as in MED. Contrary to that, rates of colon & rectum and prostate cancer are positively associated with %EP milk & products in ALL, but not in MED. The observed relationship between the risk of cancer and the consumption of food of animal origin in both European regions can explain, to a certain degree, an inconsistency revealed in epidemiological studies. Moreover the frequency of polymorphic alleles in different European regions, being an important factor in the susceptibility to carcinogens, should be taken into account in future studies, when more data on this matter will become available. It would also be interesting to find out to what extent the association between meat intake and colon & rectum, breast and prostate cancer incidence depends on meat consumption in different life periods.

The Protective Ability of Food of Plant Origin

Even low-level exposure to carcinogenic substances of more vulnerable groups may disclose the inter-individual variability in metabolic susceptibility to carcinogens, resulting in accumulation of DNA damage and leading to the genomic instability. The amount of DNA adducts can be decreased by intake of antioxidant dietary compounds. For example, fewer white blood cell DNA adducts in bladder biopsies occur among consumers of large quantities of fruits or vegetables [66]. This might be the reason why a high fruit and vegetable intake showed a preventive ability against cancer of lung, colon, esophagus, breast, prostate, stomach and heart diseases in many epidemiological studies (NAS Report, Diet and Health, 1989). Not only the abundant amount of antioxidants that enhances the scavenging capacity of cells, but also the ability of certain polyphenols and other bioactive compounds to modulate the activity of enzymes involved in carcinogen metabolism may be responsible for this effect [67–71]. In particular, the EU-funded study 'Local Food and Nutraceuticals' confirmed such an ability for more than 100 edible plants collected in the European Mediterranean region [9, 72–80].

Flavonoids are the largest group among polyphenols [81–83]. They are widely distributed in fruits and vegetables. They also inhibit pro-carcinogen bioactivation [84–88]. They also play a protective role because of their antioxidant properties [81, 89]. Consequently, flavonoids are able to decrease the amount of DNA adducts induced by HAAs, to lessen DNA damage related to

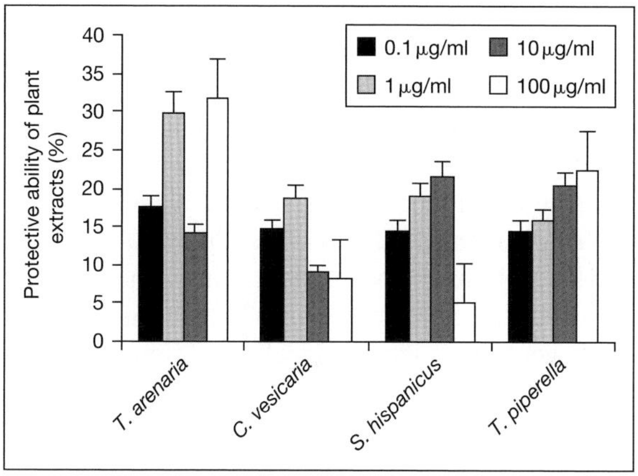

Fig. 2. Human lymphocytes were pretreated with the different concentrations of plant extracts derived from the *T. arenaria*, *C. vesicaria*, *S. hispanicus* and *T. piperella* for 1 h and then exposed to 25 μ*M* hydrogen peroxide for 5 min at 4°C. The protective ability was defined as the percentage of which the amount of oxidative DNA damage induced by hydrogen peroxide in lymphocytes is decreased due to plant extracts pretreatment (percentage of DNA in the tail was measured by comet assay).

lipid peroxidation [90, 91], and to inhibit peroxide-induced micronucleated cells [91]. Their chemopreventive activity against cancer induced by various carcinogens has been well documented in animal models [92, 93]. Furthermore, flavonoids are involved in the reduction of oxidative damage in DNA, mainly of 8-oxo-7-hydro-2′-deoxyguanosine (8-oxodG), which plays a significant role in development of various diseases, especially cancer. As 8-oxodG is one of the most common types of damage in DNA found in cancer cells, it can be used as a biomarker of carcinogenesis [94–96]. The extent of this damage can be lowered by fruit, vegetable and antioxidant supplement intake, as shown while determining oxidative DNA damage in white blood cells by comet assay [97–101]. Such a protective effect was also observed for certain edible plants investigated in the already mentioned EU Project [9, 72–80]. The dose response of certain plant extracts as *Scolymus hispanicus, Thymus piperella, Crepis vesicaria, Terfezia arenaria*, derived from local wild food plants of Southern Spain, is presented in figure 2. The association between the doses of those extracts and their DNA-protective ability against the oxidative DNA damage revealed a very broad spectrum of response, confirming the expectations that the constituent of the plant extracts is much more significant than the dose used. Since the oxidative DNA damage can play a significant role in mutagenesis, cancer, and other human pathologies, enriching the daily nutrition in antioxidant compounds seems to be the best strategy that is possible to achieve by eating food rich in antioxidants and/or by taking supplements containing polyphenols, for example plant extracts.

The preference for fresh fruits and vegetables in the Mediterranean diet often corresponds to a higher consumption of raw food in general, with more antioxidant and polyphenol intake, and lower production of cooking-related oxidants [102]. The significant differences between MED and OTHERS were also confirmed by the FAO database (table 1). However, the differences in consumption of vegetables were statistically significant: the vegetable consumption was higher in ALL and MED than in OTHERS, contrary to the statistically non-significant differences in fruit consumption between the two groups of countries. Moreover, although the consumption of vegetables increased in OTHERS in the last decade, it has not yet reached the level of MED. The lowest value was 40 g/day/person in Iceland in 1969–71 but it rose 3.5 times in the next three decades, reaching 144 g/day/person in the year 2002. In turn, the highest consumption of vegetables was 437 g/day/person in Turkey in 1969–71; that rate rose to 727 g/day/person by 2002. The mean value for MED rose from 428 g/day/person in 1969–71 to 508 g/day/person in 2000–2002, whereas in OTHERS it increased from 143 to 223 g/day/person. The average consumption for each group of countries is presented in table 1. There should be no doubt that such differences in vegetable consumption have a huge impact on health maintenance.

Many epidemiologic studies examined the relationship between fruit and vegetable intake and the incidence of different types of cancer. Findings have varied considerably [103–105]. In particular, in the US Nurses' Health Study and the Health Professionals' Follow-up Study (1980–1996), one of the largest prospective studies ever carried out (1,743,645 persons-years), no association was found for women and men between overall fruit and vegetable consumption and risk of colon and rectal cancer [106]. There were also no associations observed for subgroups of fruits and vegetables, or for individual fruits and vegetables. Therefore, although fruits and vegetables may provide protection against some chronic diseases, their frequent consumption does not appear to confer protection from colon or rectal cancer [103].

As cancer usually develops for years before it is clinically recognized and registered, the correlation between colon & rectum, breast and prostate cancer incidences recorded in 2002 and the consumption of fruits and vegetables for the previous three decades was analyzed. The relationship between fruit consumption and three types of analyzed cancers was insignificant in all analyzed groups of countries. Nonetheless, a significant reverse association between cancer incidence and the amount of vegetables and vegetable products consumed for each analyzed period (between 1969 and 2002) was observed in ALL countries (table 2). This effect did not occur in MED and OTHERS analyzed separately, although the consumption of vegetables was significantly higher in MED than in ALL and OTHERS. The average vegetable consumption in MED was 475 ± 123 g/person/day; the rate was 200 ± 48 in OTHERS and 300 ± 165 in ALL.

When the analysis was performed by grouping countries below and above the median value of vegetable consumption in each decade, instead of stratifying countries by geographical location, the significant protective ability of vegetables was visible in all decades only for prostate cancer. The amount of vegetables & products has to be >200 g/day to disclose the protective ability. The breast cancer incidences seem to depend not only on the amount of vegetable intake but also on life period. The amount of breast cancer incidences was significantly lower only in the group which consumed >220 g/day two decades before cancer was registered.

It is possible that a certain group of vegetables, which have a more protective ability against cancer than others do, is consumed in larger quantity in MED than in OTHERS.

Pulses and cereals, food of plant origin, are consumed in higher quantities in the European Mediterranean region than they are in other European countries. The FAO data show that the consumption of cereal & products (excluding beer), as well as pulses & products, was significantly higher in MED than in OTHERS and ALL, in each period analyzed (table 1). However, the average intake of cereal & products in ALL was not significantly different from that in MED and in OTHERS. When the consumption of cereals & products in each period analyzed was correlated with cancer incidence in 2002, the significant negative association was found with colon & rectum, breast and prostate cancer incidence in ALL countries, but not in OTHERS (table 2). In MED such a dependence was only found for prostate cancer (and not for colon & rectum and breast cancer). In OTHERS, the protective effect of early consuming cereals & products (two and three decades before breast cancer incidences were registered) was observed. The protective ability of pulses & products was observed for each analyzed cancer in ALL, as well as in MED, but only for breast cancer in OTHERS (table 2).

The significance of consumption of food of plant origin in cancer prevention was additionally confirmed through simple regression analysis, studying the dependence between the percentage of the energy supply from vegetable products and cancer incidences. The reverse association between this percentage of energy and colon & rectum, breast and prostate cancer incidences was statistically significant in ALL ($p = 0.001$, $p = 0.00002$, $p = 0.0001$, respectively for different cancers), but not in OTHERS. The protective effect was observed only for breast cancer in MED.

Dietary recommendations of the UK Department of Health are in line with these findings. According to those directives, for maximum benefit, one needs

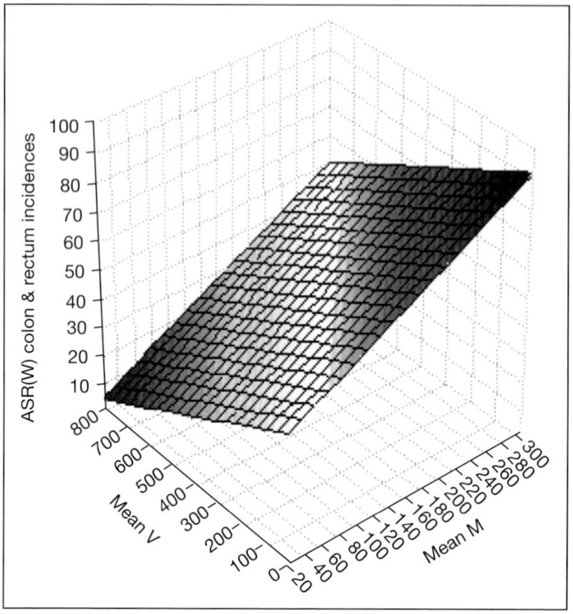

Fig. 3. A three-dimensional surface graph illustrating interaction between mean values of meat and meat product and vegetables and vegetable product consumption (g/person/day) (from 1969 to 2002) and colon & rectum cancer incidences (in 2002) in 18 European countries.

to eat at least five portions (80 g each) of various fruits and vegetables a day (http://www.dh.gov.uk/PolicyAndGuidance). This can be achieved, for example, by consuming a portion of cooked vegetables (three rounded tablespoons), half a can of canned vegetables, tomato-based cooking sauces (accompanying a portion of pasta), tomato soup, or salad (300 ml bowl of mixed leaves).

Vegetable to Meat (V/M) Ratio

The results presented in this paper clearly show that over the last three decades the colon & rectum cancer incidences have increased with growing meat consumption but diminished when vegetable intake increased. The interaction between the average consumption of meat and its products (M) and vegetables and their products (V), and colon & rectum cancer incidences is presented in a three-dimensional surface graph plot in figure 3.

As already mentioned, the intake of meat & products, as well as vegetables & products, increased from 1969 constantly but at a different rate in different

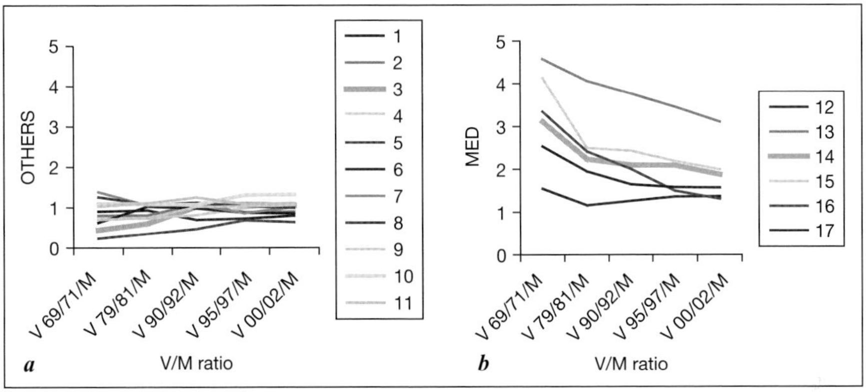

Fig. 4. Changes in ratio between vegetables and vegetable product to meat and meat product consumption during the reference period in following countries: (***a***) 1 Austria, 2 Denmark, 3 Finland, 4 Germany, 5 Iceland, 6 Ireland, 7 the Netherlands, 8 Norway, 9 Sweden, 10 Switzerland, 11 UK (referred to as OTHERS); (***b***) 12 France, 13 Greece, 14 Italy, 15 Portugal, 16 Spain, 17 Cyprus (referred to as MED). The food consumption provided by FAO refers to the amount of food for each individual per day (g).

countries. Because of that, it can be expected that the factual trend in dietary patterns in both regions of Europe over the last three decades will be better reflected by the V/M ratio. Therefore, the hypothesis about association between the V/M ratio and cancer risk should be tested.

As shown in figure 4, the V/M ratio was always >1.5 in MED (and even reached 10 in Turkey, not shown), whereas it was <1.5 in OTHERS and changed only slightly over time there. The V/M ratios drastically dropped in 1979–81 in each MED country and subsequently decreased still further, though at a slower pace. Even so, these values never decreased to the level of OTHERS. A significant correlation was found between incidence of each analyzed type of cancer and the V/M in ALL countries, as well as in MED; no such association was observed in OTHERS.

The ratio between the amount of energy from vegetables and from animal products, instead of the quantity ratios, was also used as a dietary-protective indicator. When such energy ratios were calculated for each of the 18 European countries (averages for 1961–2002) and placed on the X-axis against the colon & rectum, breast and prostate cancer incidences, a clearly defined cancer risk pattern emerged (fig. 5) Countries in which this ratio was <2 (OTHERS) had a higher cancer incidence than those in which the ratio was >2 (MED).

This confirms the hypothesis that meat and meat product carcinogenic activity can be neutralized by the consumption of the proper amount of vegetables.

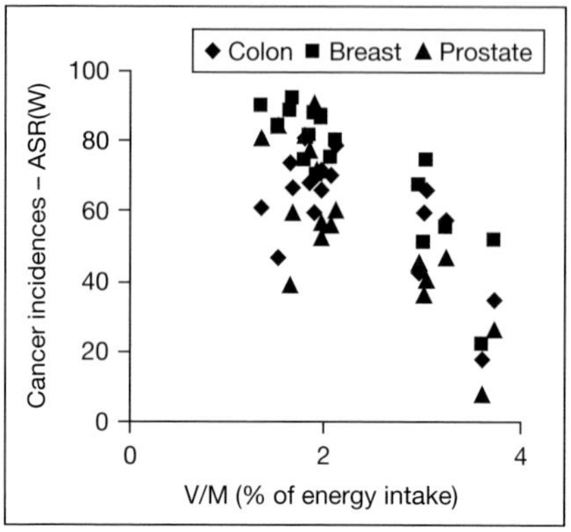

Fig. 5. Association between the ratio of the amount of energy from vegetables and animal products for each of the 18 European countries (averages for 1961–2002) and the colon & rectum, breast and prostate cancer incidences.

Final Remarks

The dietary patterns in European Mediterranean countries, reflected by the >2 ratio between consumption of vegetables & products and meat & products, diminish colon & rectum, breast and prostate cancer risk. Increased consumption of vegetables & products seems to be necessary to inactivate pro-carcinogenous substances provided by the HAAs and PAHs from fried/grilled meat. When more toxic substances are consumed, more antioxidants and polyphenols are necessary for detoxifications. In short, the protective nutritional environment for DNA molecules seems to be essential to maintain the normal functioning of cells and prevent their transformation (fig. 1). Since there is no available data for European regions on frequencies of polymorphisms in genes involved in detoxifying reactions, it is reasonable to assume that the V/M ratio should be even higher in places where detoxifying enzymes have a lower activity.

Our genetic constitution was established at a time when the environment was quite different from what it is today, and that constitution has changed very little during the last 10,000 years. It is consequently very tempting to consider the Late Paleolithic diet as a very beneficial pattern to follow. The proportion between the constituencies of food in that diet seems to be especially important. As many publications analyzing the evolutionary aspect of diet have highlighted, humans' contemporary diet, as compared to Paleolithic one, is characterized by excessive energy intake relative to energy expenditure, lower complex carbohydrates, less fiber, more fat (in particular saturated fatty acids), and the presence of trans fatty acids, unknown to our ancestors. All these characteristics may contribute to a higher risk of many degenerative diseases. Most important, however, seems to be the drastic decrease in the consumption of food originating from plants; this decrease causes the DNA damage to remain at a higher level. A low level of protective plant compounds in the diet cannot adequately neutralize the reactive oxygen species and also cannot support the enzymatic system in our organism in sufficient removal of DNA damage, especially in the situation when cells are attacked by the enhanced amount of toxic substances present in 'unhealthy' food products.

Through the reconstruction of prehistoric dietary patterns from anthropological evidence, and the observation of contemporary hunter-gatherer societies, it is known that prehistoric diets provided abundant amounts of proteins from wild game, fish and carbohydrates from uncultivated plants (characterized by a high quantity of fiber), and had a different fat profile [107, 108]. From a simple model of the Late Paleolithic human diet, which assumed a 35:65 animal/plant subsistence ratio and a daily average intake of 3,000 kcal, and also taking into account the nutritional values for game (1.42 kcal/g) and wild plant (1.29 kcal/g) food, it was calculated that the amount of food of plant origin eaten daily by our ancestors was about 1,400–1,500 g, and of wild game was about 788 g [109–111]. These figures imply a V/M ratio of 1.9. This ratio is close to the one calculated above that is linked to a lower cancer risk in the European Mediterranean region in spite of the commercial changes that have occurred in meat and vegetable products produced for consumption. The recognition of the importance of the ratio between food constituencies for human health, as has occurred already for unsaturated fatty acids [112–115], can facilitate the development of more precise dietary recommendations involving other compounds.

The results in the present paper provide additional evidence for the beneficial health effect of the current Mediterranean diet. However, in the last two decades the food pattern of the Mediterranean population was subjected to substantial changes, particularly in regard to the ratio between the amount of vegetables and meat consumed, which in turn resulted in a decrease of the amount of biologically active polyphenols and antioxidant vitamins, lowering the defense ability of the body against genotoxic substances. The significant drop of the V/M ratio may cause that the number of cancer incidences in this region of Europe will rise in the near future.

It is not very likely that consumption of fruits and vegetables will reach a level required for neutralizing the negative effect of meat intake. Therefore, the modern nutritional strategies propose inclusion of nutraceuticals rich in antioxidant compounds into the diet. Taking dietary supplements, particularly plant extracts, is beginning to be believed as bringing health benefits quicker than would normally be the case through eating conventionally healthy food alone. Studies identifying edible plants having certain health beneficial properties, as those investigated in the EU Project 'Local Food and Nutraceuticals', significantly contribute to a better understanding of the problem.

Acknowledgements

This project was supported by a grant from the European Commission (Local Food-Nutraceuticals, QLK-2001-00173) and by the State Committee for Scientific Research, Warsaw, Poland, Project No. 3 P05E 016 25. I sincerely thank Prof. Aleksander Koj for helpful discussion and support and Ewa Soltys and Grzegorz Zajac for performing the experiments with the plant extracts.

References

1 Giacosa A: The Mediterranean diet and its protective role against cancer. Eur J Cancer Prev 2004;13:155–157.
2 Giacosa A, Filiberti R, Visconti P, Puntoni R: Mediterranean diet and digestive precancerous lesions. Eur J Cancer Prev 1993;2(suppl 2):17–26.
3 Simopoulos AP: The Mediterranean diets: what is so special about the diet of Greece? The scientific evidence. J Nutr 2001;131(suppl):3065S–3073S.
4 Simopoulos AP, Sidossis LS: What is so special about the traditional diet of Greece. The scientific evidence. World Rev Nutr Diet 2000;87:24–42.
5 Trichopoulos D, Lagiou P: Mediterranean diet and overall mortality differences in the European Union. Public Health Nutr 2004;7:949–951.
6 Trichopoulou A: Traditional Mediterranean diet and longevity in the elderly: a review. Public Health Nutr 2004;7:943–947.
7 Trichopoulou A, Naska A, Vasilopoulou E: Guidelines for the intake of vegetables and fruit: the Mediterranean approach. Int J Vitam Nutr Res 2001;71:149–153.
8 Rumm-Kreuter D: Comparison of the eating and cooking habits of northern Europe and the Mediterranean countries in the past, present and future. Int J Vitam Nutr Res 2001;71:141–148.
9 Heinrich M, Leonti M, Nebel S, Peschel W: 'Local Food-Nutraceuticals': an example of a multidisciplinary research project on local knowledge. J Physiol Pharmacol 2005;56(suppl 1):5–22.
10 Visioli F, Bogani P, Grande S, Galli C: Mediterranean food and health: building human evidence. J Physiol Pharmacol 2005;56(suppl 1):37–49.
11 Trichopoulou A, Bamia C, Trichopoulos D: Mediterranean diet and survival among patients with coronary heart disease in Greece. Arch Intern Med 2005;165:929–935.
12 Gill CI, Boyd A, McDermott E, McCann M, Servili M, Selvaggini R, Taticchi A, Esposto S, Montedoro G, McGlynn H, et al: Potential anti-cancer effects of virgin olive oil phenols on colorectal carcinogenesis models in vitro. Int J Cancer 2005;117:1–7.
13 Cottet V, Bonithon-Kopp C, Kronborg O, Santos L, Andreatta R, Boutron-Ruault MC, Faivre J: Dietary patterns and the risk of colorectal adenoma recurrence in a European intervention trial. Eur J Cancer Prev 2005;14:21–29.
14 Visioli F, Grande S, Bogani P, Galli C: The role of antioxidants in the Mediterranean diets: focus on cancer. Eur J Cancer Prev 2004;13:337–343.
15 La Vecchia C: Mediterranean diet and cancer. Public Health Nutr 2004;7:965–968.
16 Simopoulos AP: The traditional diet of Greece and cancer. Eur J Cancer Prev 2004;13:219–230.
17 Galli C, Visioli F: Antioxidant properties of Mediterranean diet. Int J Vitam Nutr Res 2001;71: 185–188.
18 Mathers JC: The biological revolution – towards a mechanistic understanding of the impact of diet on cancer risk. Mutat Res 2004;551:43–49.
19 Wiseman H, Kaur H, Halliwell B: DNA damage and cancer: measurement and mechanism. Cancer Lett 1995;93:113–120.
20 Dixon LB, Balder HF, Virtanen MJ, Rashidkhani B, Mannisto S, Krogh V, van den Brandt PA, Hartman AM, Pietinen P, Tan F, et al: Dietary patterns associated with colon and rectal cancer: results from the Dietary Patterns and Cancer (DIETSCAN) Project. Am J Clin Nutr 2004;80: 1003–1011.

21 Fung T, Hu FB, Fuchs C, Giovannucci E, Hunter DJ, Stampfer MJ, Colditz GA, Willett WC: Major dietary patterns and the risk of colorectal cancer in women. Arch Intern Med 2003;163:309–314.
22 Singh PN, Fraser GE: Dietary risk factors for colon cancer in a low-risk population. Am J Epidemiol 1998;148:761–774.
23 Nkondjock A, Ghadirian P: Associated nutritional risk of breast and colon cancers: a population-based case-control study in Montreal, Canada. Cancer Lett 2005;223:85–91.
24 Kim MK, Sasaki S, Otani T, Tsugane S: Dietary patterns and subsequent colorectal cancer risk by subsite: a prospective cohort study. Int J Cancer 2005;115:790–798.
25 Wu K, Hu FB, Fuchs C, Rimm EB, Willett WC, Giovannucci E: Dietary patterns and risk of colon cancer and adenoma in a cohort of men (United States). Cancer Causes Control 2004;15:853–862.
26 Gallus S, Bosetti C, La Vecchia C: Mediterranean diet and cancer risk. Eur J Cancer Prev 2004;13: 447–452.
27 Cross AJ, Sinha R: Meat-related mutagens/carcinogens in the etiology of colorectal cancer. Environ Mol Mutagen 2004;44:44–55.
28 Marnett LJ, Plastaras JP: Endogenous DNA damage and mutation. Trends Genet 2001;17: 214–221.
29 Quiles JL, Ochoa JJ, Ramirez-Tortosa C, Battino M, Huertas JR, Martin Y, Mataix J: Dietary fat type (virgin olive vs. sunflower oils) affects age-related changes in DNA double-strand breaks, antioxidant capacity and blood lipids in rats. Exp Gerontol 2004;39:1189–1198.
30 Vogel U, Danesvar B, Autrup H, Risom L, Weimann A, Poulsen HE, Moller P, Loft S, Wallin H, Dragsted LO: Effect of increased intake of dietary animal fat and fat energy on oxidative damage, mutation frequency, DNA adduct level and DNA repair in rat colon and liver. Free Radic Res 2003;37:947–956.
31 Kikugawa K, Yasuhara Y, Ando K, Koyama K, Hiramoto K, Suzuki M: Effect of supplementation of n–3 polyunsaturated fatty acids on oxidative stress-induced DNA damage of rat hepatocytes. Biol Pharm Bull 2003;26:1239–1244.
32 Snyderwine EG, Schut HA, Sugimura T, Nagao M, Adamson RH: DNA adduct levels of 2-amino-1-methyl-6-phenylimidazo-[4,5-b]pyridine (PhIP) in tissues of cynomolgus monkeys after single or multiple dosing. Carcinogenesis 1994;15:2757–2761.
33 Ghoshal A, Preisegger KH, Takayama S, Thorgeirsson SS, Snyderwine EG: Induction of mammary tumors in female Sprague-Dawley rats by the food-derived carcinogen 2-amino-1-methyl-6-phenylimidazo[4,5-b]pyridine and effect of dietary fat. Carcinogenesis 1994;15:2429–2433.
34 Weisburger JH: Comments on the history and importance of aromatic and heterocyclic amines in public health. Mutat Res 2002;506–507:9–20.
35 Spingarn NE, Kasai H, Vuolo LL, Nishimura S, Yamaizumi Z, Sugimura T, Matsushima T, Weisburger JH: Formation of mutagens in cooked foods. III. Isolation of a potent mutagen from beef. Cancer Lett 1980;9:177–183.
36 Felton JS, Knize MG, Shen NH, Andresen BD, Bjeldanes LF, Hatch FT: Identification of the mutagens in cooked beef. Environ Health Perspect 1986;67:17–24.
37 Felton JS, Knize MG, Shen NH, Lewis PR, Andresen BD, Happe J, Hatch FT: The isolation and identification of a new mutagen from fried ground beef: 2-amino-1-methyl-6-phenylimidazo [4,5-b]pyridine (PhIP). Carcinogenesis 1986;7:1081–1086.
38 Bingham SA: Diet and colorectal cancer prevention. Biochem Soc Trans 2000;28:12–16.
39 Talalay P, Prochaska HJ, Spencer SR: Regulation of enzymes that detoxify the electrophilic forms of chemical carcinogens. Princess Takamatsu Symp 1990;21:177–187.
40 Mattano SS, Weber WW: Kinetics of arylamine *N*-acetyltransferase in tissues from rapid and slow acetylator mice. Carcinogenesis 1987;8:133–137.
41 Hein DW: *N*-Acetyltransferase genetics and their role in predisposition to aromatic and heterocyclic amine-induced carcinogenesis. Toxicol Lett 2000;112/113:349–356.
42 Hein DW, Doll MA, Fretland AJ, Leff MA, Webb SJ, Xiao GH, Devanaboyina US, Nangju NA, Feng Y: Molecular genetics and epidemiology of the NAT1 and NAT2 acetylation polymorphisms. Cancer Epidemiol Biomarkers Prev 2000;9:29–42.
43 Bell DA, Badawi AF, Lang NP, Ilett KF, Kadlubar FF, Hirvonen A: Polymorphism in the *N*-acetyltransferase 1 (NAT1) polyadenylation signal: association of NAT1*10 allele with higher *N*-acetylation activity in bladder and colon tissue. Cancer Res 1995;55:5226–5229.

44 Bell DA, Stephens EA, Castranio T, Umbach DM, Watson M, Deakin M, Elder J, Hendrickse C, Duncan H, Strange RC: Polyadenylation polymorphism in the acetyltransferase-1 gene (NAT1) increases risk of colorectal cancer. Cancer Res 1995;55:3537–3542.

45 Butcher NJ, Ilett KF, Minchin RF: Substrate-dependent regulation of human arylamine *N*-acetyltransferase-1 in cultured cells. Mol Pharmacol 2000;57:468–473.

46 Wadelius M, Andersson AO, Johansson JE, Wadelius C, Rane E: Prostate cancer associated with CYP17 genotype. Pharmacogenetics 1999;9:635–639.

47 Wadelius M, Autrup JL, Stubbins MJ, Andersson SO, Johansson JE, Wadelius C, Wolf CR, Autrup H, Rane A: Polymorphisms in NAT2, CYP2D6, CYP2C19 and GSTP1 and their association with prostate cancer. Pharmacogenetics 1999;9:333–340.

48 Dipple A: DNA adducts of chemical carcinogens. Carcinogenesis 1995;16:437–441.

49 Ghoshal A, Davis CD, Schut HA, Snyderwine EG: Possible mechanisms for PhIP-DNA adduct formation in the mammary gland of female Sprague-Dawley rats. Carcinogenesis 1995;16: 2725–2731.

50 Pfau W, O'Hare MJ, Grover PL, Phillips DH: Metabolic activation of the food mutagens 2-amino-3-methylimidazo[4,5-f]quinoline (IQ) and 2-amino-3,4-dimethylimidazo[4,5-f]quinoline (MeIQ) to DNA binding species in human mammary epithelial cells. Carcinogenesis 1992;13:907–909.

51 Pfau W, Stone EM, Brockstedt U, Carmichael PL, Marquardt H, Phillips DH: DNA adducts in human breast tissue: association with *N*-acetyltransferase-2 (NAT2) and NAT1 genotypes. Cancer Epidemiol Biomarkers Prev 1998;7:1019–1025.

52 Badawi AF, Hirvonen A, Bell DA, Lang NP, Kadlubar FF: Role of aromatic amine acetyltransferases, NAT1 and NAT2, in carcinogen-DNA adduct formation in the human urinary bladder. Cancer Res 1995;55:5230–5237.

53 Badawi AF, Stern SJ, Lang NP, Kadlubar FF: Cytochrome P_{450} and acetyltransferase expression as biomarkers of carcinogen-DNA adduct levels and human cancer susceptibility. Prog Clin Biol Res 1996;395:109–140.

54 Butler MA, Lang NP, Young JF, Caporaso NE, Vineis P, Hayes RB, Teitel CH, Massengill JP, Lawsen MF, Kadlubar FF: Determination of CYP1A2 and NAT2 phenotypes in human populations by analysis of caffeine urinary metabolites. Pharmacogenetics 1992;2:116–127.

55 Murtaugh MA, Ma KN, Sweeney C, Caan BJ, Slattery ML: Meat consumption patterns and preparation, genetic variants of metabolic enzymes, and their association with rectal cancer in men and women. J Nutr 2004;134:776–784.

56 Murtaugh MA, Sweeney C, Ma KN, Caan BJ, Slattery ML: The CYP1A1 genotype may alter the association of meat consumption patterns and preparation with the risk of colorectal cancer in men and women. J Nutr 2005, 135:179–186.

57 Hemminki K, Vaittinen P: Familial cancers in a nationwide family cancer database: age distribution and prevalence. Eur J Cancer 1999;35:1109–1117.

58 Hemminki K, Vaittinen P: Familial risks in second primary breast cancer based on a family cancer database. Eur J Cancer 1999;35:455–458.

59 Alavanja MC, Brownson RC, Benichou J: Estimating the effect of dietary fat on the risk of lung cancer in nonsmoking women. Lung Cancer 1996;14(suppl 1):S63–S74.

60 Breuer-Katschinski B, Nemes K, Marr A, Rump B, Leiendecker B, Breuer N, Goebell H: Colorectal adenomas and diet: a case-control study. Colorectal Adenoma Study Group. Dig Dis Sci 2001;46:86–95.

61 Boyd NF, Stone J, Vogt KN, Connelly BS, Martin LJ, Minkin S: Dietary fat and breast cancer risk revisited: a meta-analysis of the published literature. Br J Cancer 2003;89:1672–1685.

62 Skog K, Solyakov A, Arvidsson P, Jagerstad M: Analysis of nonpolar heterocyclic amines in cooked foods and meat extracts using gas chromatography-mass spectrometry. J Chromatogr A 1998;803:227–233.

63 Skog KI, Johansson MA, Jagerstad MI: Carcinogenic heterocyclic amines in model systems and cooked foods: a review on formation, occurrence and intake. Food Chem Toxicol 1998;36: 879–896.

64 Giovannucci E, Goldin B: The role of fat, fatty acids, and total energy intake in the etiology of human colon cancer. Am J Clin Nutr 1997;66(suppl):1564S–1571S.

65 Fleshner N, Bagnell PS, Klotz L, Venkateswaran V: Dietary fat and prostate cancer. J Urol 2004;171:S19–S24.
66 Peluso M, Vineis P: 'Mediterranean diet' and DNA adducts. Eur J Cancer Prev 2000;9:71–72.
67 Vineis P, Malats N, Boffetta P: Why study metabolic susceptibility to cancer? IARC Sci Publ 1999;148:1–3.
68 Vineis P, Malats N, Porta M, Real FX: Human cancer, carcinogenic exposures and mutation spectra. Mutat Res 1999;436:185–194.
69 Schwarz D, Roots I: In vitro assessment of inhibition by natural polyphenols of metabolic activation of procarcinogens by human CYP1A1. Biochem Biophys Res Commun 2003;303:902–907.
70 Wilkinson JT, Clapper ML: Detoxication enzymes and chemoprevention. Proc Soc Exp Biol Med 1997;216:192–200.
71 Visioli F: Antioxidants in Mediterranean diets. World Rev Nutr Diet 2000;87:43–55.
72 Rivera D, Obon C, Inocencio C, Heinrich M, Verde A, Fajardo J, Llorach R: The ethnobotanical study of local Mediterranean food plants as medicinal resources in Southern Spain. J Physiol Pharmacol 2005;56(suppl 1):97–114.
73 Dulak J: Nutraceuticals as antiangiogenic agents: hopes and reality. J Physiol Pharmacol 2005;56 (suppl 1):51–69.
74 Kapiszewska M, Soltys E, Visioli F, Cierniak A, Zajac G: The protective ability of the Mediterranean plant extracts against the oxidative DNA damage. The role of the radical oxygen species and the polyphenol content. J Physiol Pharmacol 2005;56(suppl 1):183–197.
75 Kapiszewska M, Zajac G, Kalemba M, Soltys E: The estrogenic status and the COMT genotype of female blood donors influence the protective ability of the Mediterranean plant extracts against the hydrogen peroxide-induced DNA damage in lymphocytes. J Physiol Pharmacol 2005;56(suppl 1):199–217.
76 Manios Y, Antonopoulou S, Kaliora AC, Felliou G, Perrea D: Dietary intake and biochemical risk factors for cardiovascular disease in two rural regions of Crete. J Physiol Pharmacol 2005;56 (suppl 1):171–181.
77 Schaffer S, Schmitt-Schillig S, Muller WE, Eckert GP: Antioxidant properties of Mediterranean food plant extracts: geographical differences. J Physiol Pharmacol 2005;56(suppl 1):115–124.
78 Schmitt-Schillig S, Schaffer S, Weber CC, Eckert GP, Muller WE: Flavonoids and the aging brain. J Physiol Pharmacol 2005;56(suppl 1):23–36.
79 Stalinska K, Guzdek A, Rokicki M, Koj A: Transcription factors as targets of the anti-inflammatory treatment. A cell culture study with extracts from some Mediterranean diet plants. J Physiol Pharmacol 2005;56(suppl 1):157–169.
80 Strzelecka M, Bzowska M, Koziel J, Szuba B, Dubiel O, Riviera Nunez D, Heinrich M, Bereta J: Anti-inflammatory effects of extracts from some traditional Mediterranean diet plants. J Physiol Pharmacol 2005;56(suppl 1):139–156.
81 Clifford MN: Diet-derived phenols in plasma and tissues and their implications for health. Planta Med 2004;70:1103–1114.
82 Tutel'ian VA, Baturin AK, Martinchik EA: Flavonoids: food content, consumption, bioavailability (in Russian). Vopr Pitan 2004;73:43–48.
83 Yao LH, Jiang YM, Shi J, Tomas-Barberan FA, Datta N, Singanusong R, Chen SS: Flavonoids in food and their health benefits. Plant Foods Hum Nutr 2004;59:113–122.
84 Chen S, Zhou D, Okubo T, Kao YC, Eng ET, Grube B, Kwon A, Yang C, Yu B: Prevention and treatment of breast cancer by suppressing aromatase activity and expression. Ann NY Acad Sci 2002;963:229–238.
85 Dingley KH, Ubick EA, Chiarappa-Zucca ML, Nowell S, Abel S, Ebeler SE, Mitchell AE, Burns SA, Steinberg FM, Clifford AJ: Effect of dietary constituents with chemopreventive potential on adduct formation of a low dose of the heterocyclic amines PhIP and IQ and phase II hepatic enzymes. Nutr Cancer 2003;46:212–221.
86 Scharf G, Prustomersky S, Knasmuller S, Schulte-Hermann R, Huber WW: Enhancement of glutathione and γ-glutamylcysteine synthetase, the rate-limiting enzyme of glutathione synthesis, by chemoprotective plant-derived food and beverage components in the human hepatoma cell line HepG2. Nutr Cancer 2003;45:74–83.

87 Schwarz D, Kisselev P, Roots I: CYP1A1 genotype-selective inhibition of benzo[a]pyrene activation by quercetin. Eur J Cancer 2005;41:151–158.
88 Tate P, God J, Bibb R, Lu Q, Larcom LL: Inhibition of metalloproteinase activity by fruit extracts. Cancer Lett 2004;212:153–158.
89 Leighton F, Cuevas A, Guasch V, Perez DD, Strobel P, San Martin A, Urzua U, Diez MS, Foncea R, Castillo O, et al: Plasma polyphenols and antioxidants, oxidative DNA damage and endothelial function in a diet and wine intervention study in humans. Drugs Exp Clin Res 1999;25:133–141.
90 Sanchez-Perez Y, Carrasco-Legleu C, Garcia-Cuellar C, Perez-Carreon J, Hernandez-Garcia S, Salcido-Neyoy M, Aleman-Lazarini L, Villa-Trevino S: Oxidative stress in carcinogenesis. Correlation between lipid peroxidation and induction of preneoplastic lesions in rat hepatocarcinogenesis. Cancer Lett 2005;217:25–32.
91 Fenech M, Stockley C, Aitken C: Moderate wine consumption protects against hydrogen peroxide-induced DNA damage. Mutagenesis 1997;12:289–296.
92 Prasad L, Khan TH, Sehrawat A, Sultana S: Modulatory effect of *Morus indica* against two-stage skin carcinogenesis in Swiss albino mice: possible mechanism by inhibiting aryl hydrocarbon hydroxylase. J Pharm Pharmacol 2004;56:1291–1298.
93 Ganguly C, De S, Das S: Prevention of carcinogen-induced mouse skin papilloma by whole fruit aqueous extract of *Momordica charantia*. Eur J Cancer Prev 2000;9:283–288.
94 Breton J, Sichel F, Pottier D, Prevost V: Measurement of 8-oxo-7,8-dihydro-2′-deoxyguanosine in peripheral blood mononuclear cells: optimisation and application to samples from a case-control study on cancers of the oesophagus and cardia. Free Radic Res 2005;39:21–30.
95 Takabayashi F, Tahara S, Kaneko T, Miyoshi Y, Harada N: Accumulation of 8-oxo-2′-deoxyguanosine (as a biomarker of oxidative DNA damage) in the tissues of aged hamsters and change in antioxidant enzyme activities after single administration of *N*-nitrosobis(2-oxopropyl) amine. Gerontology 2004;50:57–63.
96 Sharma RA, Farmer PB: Biological relevance of adduct detection to the chemoprevention of cancer. Clin Cancer Res 2004;10:4901–4912.
97 Ferguson LR, Philpott M, Karunasinghe N: Dietary cancer and prevention using antimutagens. Toxicology 2004;198:147–159.
98 Djuric Z, Depper JB, Uhley V, Smith D, Lababidi S, Martino S, Heilbrun LK: Oxidative DNA damage levels in blood from women at high risk for breast cancer are associated with dietary intakes of meats, vegetables, and fruits. J Am Diet Assoc 1998;98:524–528.
99 Collins AR, Harrington V, Drew J, Melvin R: Nutritional modulation of DNA repair in a human intervention study. Carcinogenesis 2003;24:511–515.
100 Collins AR, Olmedilla B, Southon S, Granado F, Duthie SJ: Serum carotenoids and oxidative DNA damage in human lymphocytes. Carcinogenesis 1998;19:2159–2162.
101 Giovannelli L, Saieva C, Masala G, Testa G, Salvini S, Pitozzi V, Riboli E, Dolara P, Palli D: Nutritional and lifestyle determinants of DNA oxidative damage: a study in a Mediterranean population. Carcinogenesis 2002;23:1483–1489.
102 Ghiselli A, D'Amicis A, Giacosa A: The antioxidant potential of the Mediterranean diet. Eur J Cancer Prev 1997;(suppl 1):S15–S19.
103 Michels KB, Edward G, Joshipura KJ, Rosner BA, Stampfer MJ, Fuchs CS, Colditz GA, Speizer FE, Willett WC: Prospective study of fruit and vegetable consumption and incidence of colon and rectal cancers. J Natl Cancer Inst 2000;92:1740–1752.
104 Sandler RS: Epidemiology and risk factors for colorectal cancer. Gastroenterol Clin North Am 1996;25:717–735.
105 Snyderwine EG: Diet and mammary gland carcinogenesis. Recent Results Cancer Res 1998;152:3–10.
106 Linseisen J, Kesse E, Slimani N, Bueno-De-Mesquita HB, Ocke MC, Skeie G, Kumle M, Dorronsoro Iraeta M, Morote Gomez P, Janzon L, et al: Meat consumption in the European Prospective Investigation into Cancer and Nutrition (EPIC) cohorts: results from 24-hour dietary recalls. Public Health Nutr 2002;5:1243–1258.
107 Cordain L, Eaton SB, Sebastian A, Mann N, Lindeberg S, Watkins BA, O'Keefe JH, Brand-Miller J: Origins and evolution of the Western diet: health implications for the 21st century. Am J Clin Nutr 2005;81:341–354.

108 Cordain L, Eaton SB, Miller JB, Mann N, Hill K: The paradoxical nature of hunter-gatherer diets: meat-based, yet non-atherogenic. Eur J Clin Nutr 2002;56(suppl 1):S42–S52.
109 Eaton SB, Cordain L, Eaton SB: An evolutionary foundation for health promotion. World Rev Nutr Diet 2001;90:5–12.
110 Eaton SB, Eaton SB 3rd: Paleolithic vs. modern diets – selected pathophysiological implications. Eur J Nutr 2000;39:67–70.
111 Cordain L, Miller JB, Eaton SB, Mann N, Holt SH, Speth JD: Plant-animal subsistence ratios and macronutrient energy estimations in worldwide hunter-gatherer diets. Am J Clin Nutr 2000;71: 682–692.
112 Crawford M, Galli C, Visioli F, Renaud S, Simopoulos AP, Spector AA: Role of plant-derived omega–3 fatty acids in human nutrition. Ann Nutr Metab 2000;44:263–265.
113 Galli C, Visioli F: n–3 fatty acids and antioxidants in coronary heart disease. Ital Heart J 2000;1 (suppl 3):S20–S21.
114 Cuco G, Fernandez-Ballart J, Marti-Henneberg C, Arija V: The contribution of foods to the dietary lipid profile of a Spanish population. Public Health Nutr 2002;5:747–755.
115 Krajcovicova-Kudlackova M, Simoncic R, Bederova A, Klvanova J: Plasma fatty acid profile and alternative nutrition. Ann Nutr Metab 1997;41:365–370.

Maria Kapiszewska
Department of General Biochemistry
Faculty of Biotechnology, Jagiellonian University
Gronostajowa 7, PL–30-387 Kraków (Poland)
Tel. +48 12 2622174, Fax +48 12 6646919, E-Mail mkapisz@if.uj.edu.pl

Heinrich M, Müller WE, Galli C (eds): Local Mediterranean Food Plants and Nutraceuticals. Forum Nutr. Basel, Karger, 2006, vol 59, pp 154–170

Mediterranean Diet as a Nutrition Education and Dietary Guide: Misconceptions and the Neglected Role of Locally Consumed Foods and Wild Green Plants

Yannis Manios[a], *Vivian Detopoulou*[a], *Francesco Visioli*[b], *Claudio Galli*[b]

[a]Department of Nutrition and Dietetics, Harokopio University, Athens, Greece;
[b]Department of Pharmacological Sciences, University of Milan, Milan, Italy

Abstract

In the middle of the previous century the Seven Countries Study first revealed the health benefits of the traditional Cretan diet. The Cretan diet was subsequently used as a basis to form the worldwide known 'Mediterranean diet'. This dietary scheme was visualized as a food pyramid, aimed to constitute a nutrition education tool and guide for the general public and scientific community. However, the way this dietary guide has been perceived by both the public and in certain cases by the scientific community may be oversimplified. From the nutritional point of view, some of the neglected parts of this diet concern the role of locally consumed wild greens, herbs, walnuts, figs and snails, all sources of n–3 fatty acids. The above foods with the addition of fish provide a n–6:n–3 ratio of 2:1 whereas in Northern Europe and the USA the same ratio is 10–20:1. Moreover, the flavonoid and antioxidant content of the traditional Cretan diet may have been underestimated. Despite the increasing knowledge on the bioprotective profile of the traditional Cretan diet, there is a need to revisit the way this knowledge is transferred to the public emphasizing the importance of some neglected food items and nutrients.

Worldwide studies have highlighted the health benefits of certain dietary schemes [1–7]. This is the case of the Mediterranean diet, which has been proposed to be the new 'gold standard' for the prevention of chronic diseases [8, 9]. The Mediterranean diet as a 'whole diet approach' in conjunction with data on foods and nutrients, has been used as a prototype for the formation of dietary guidelines [10, 11]. Still, it is believed that several components of this dietary scheme have been underestimated, neglected or not properly underlined within the existing guidelines. The purpose of the current review is to provide a critical definition of the Mediterranean diet and present the contributory role of locally consumed foods and certain food groups in the makeup of the bioprotective profile of the Mediterranean diet.

There Are Many Mediterranean Diets

The general term *Mediterranean diet* inevitably implies a common dietary scheme consumed by the countries of the Mediterranean basin. However, this is not the case since the diets of the populations around the Mediterranean Sea present great divergences. In fact, differences in dietary patterns are evident in parts of Italy, Greece, France, Portugal, Spain, North Africa and the Middle East [12–14]. This dietary polymorphism partially reflects religious and cultural differences. Muslims for example are not allowed to consume either pork or alcoholic drinks ('Haram' foods). Wine consumption, however, is an inherent feature of other Mediterranean diets, with the striking example of the French diet and its attached 'French paradox' [15]. In parallel, the Greek Orthodox tradition includes in its rituals the consumption of wine and recommends abstention from meat and dairy products as part of fasting which can cover large periods throughout the year [16, 17]. Another considerable difference is seen between Greece and Italy. In Greece, the fat intake (mainly coming from olive oil) amounts to 40% of the total energy intake whereas in Italy the fat intake (again primarily coming from olive oil) accounts for 30% of the energy intake. Furthermore, the Greek diet is characterized by the consumption of whole grain bread and salads, whereas the main component of the Italian Mediterranean diet variant is the high consumption of pasta. As far as the Spanish diet is concerned, fish holds a special position, which is not seen to such an extent in the rest of the Mediterranean countries [18].

It is noteworthy, however, that even within the same country, significant dietary differences can be seen. In Italy the consumption of cereals, fruit and vegetables is higher in the southern part of the country whereas milk and dairy products consumption is higher in the northern part [19, 20]. Differences are also observed in Greece, as far as the fat intake pattern is concerned. On the island of Crete, fat provided approximately 40% of energy and on the island of Corfu it yielded 31–34% of energy according to food tables [21]. The difficulty in describing successfully the Mediterranean diet was underlined by remarking that '... The all embracing term "Mediterranean diet" should not be used in scientific literature until its composition, both in foods, nutrients and non-nutrients is more clearly defined and the metabolic basis of its health-promoting virtues

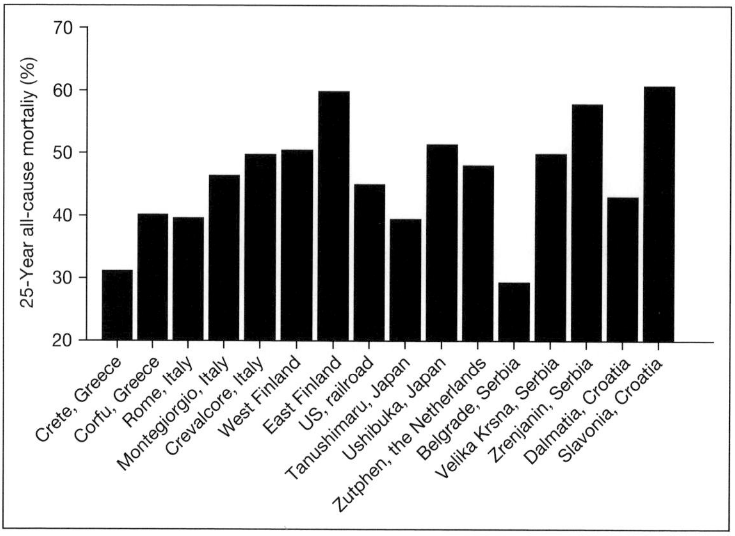

Fig. 1. Mortality from all causes in the 25-year follow-up of the Seven Countries Study.

has been better explained' [22]. Furthermore, as Simopoulos and Visioli [23] suggest, there is not just one Mediterranean diet but many Mediterranean diets and the plausible question that arises is: Which Mediterranean diet has the most bioprotective profile?

The Benefit and Food Content of the Traditional Cretan Diet

Among the variants of the Mediterranean diet, special attention has been given to the diet of Crete, an island of Greece, in which dietary patterns were relatively uniform over a 4,000-year period until the middle of the previous century [24]. At that time the mortality rate in Crete was found to be the lowest among the Mediterranean countries [24]. The virtues and the comparative superiority of the traditional Cretan diet were revealed a few years later by the Seven Countries Study, launched by Keys [25] in the 1950s. In this epidemiological study, the cohort of Crete, followed by that of Japan, presented the lowest rates of cardiovascular disease and cancer. It is of particular interest that after 25 years of follow-up, Cretans continued to have a very low cumulative mortality from coronary heart disease, cancer as well as from all causes [26–28], among the 16 cohorts, including 9 Mediterranean cohorts from Southern Europe (figs. 1–3). The Seven Countries Study accentuated the scientific interest, as it was thought that 'Cretans must be doing – or eating – something right' [29].

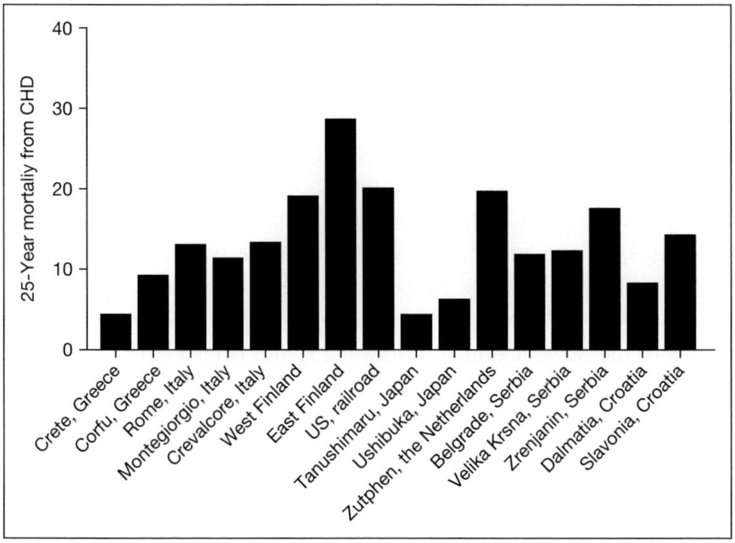

Fig. 2. Mortality from cardiovascular heart disease (CHD) in the 25-year follow-up of the Seven Countries Study.

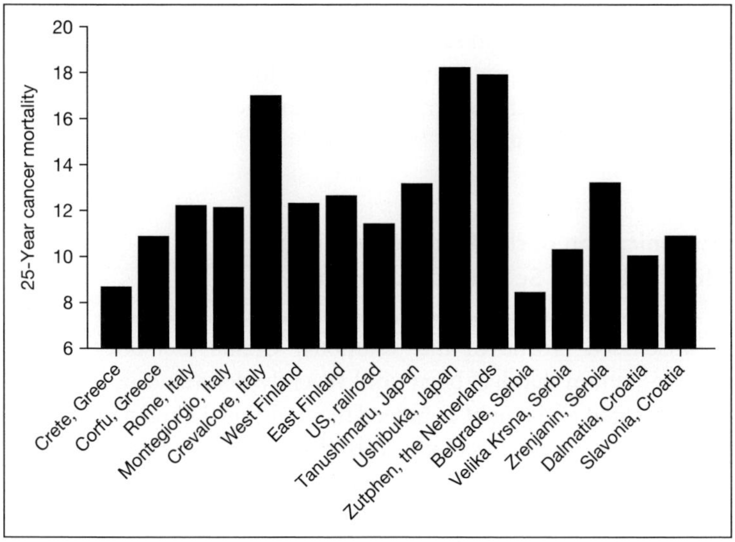

Fig. 3. Mortality from cancer in the 25-year follow-up of the Seven Countries Study.

What did Cretans eat in the middle of the previous century to achieve such health benefits? In 1948 the traditional Cretan diet was described as 'olives, cereals, grains, pulses, wild greens, herbs and fruits, together with limited quantities of goat meat and milk and fish . . . no meal was completed without bread . . . olives and olive oil contributed heavily to the energy intake . . . food seemed to be "swimming" in oil' [30]. Moreover, the traditional Cretan diet contained fish and octopus, reflecting Crete's proximity to the sea [13], pies made of a variety of wild greens [31], eggs, onions, garlic, snails, figs, walnuts, wine and other alcoholic drinks like raki [32–34]. In table 1 a detailed weekly diet is presented, constructed retrospectively by Kafatos et al. [32] from 7-day records of several Cretan participants of the Seven Countries Study [25] and from a study of Cretan men conducted in 1965 [35].

As Simopoulos emphasizes [33], wine, fruits, vegetables and olive oil, which were central components of the traditional Cretan diet, contained resveratrol, glutathione, vitamin C, vitamin E, lycopene, β-carotene, polyphenols and other antioxidants. Important contributors to the antioxidant properties of the Mediterranean diet were also tomatoes, onions, garlic and herbs, which provided carotenoids, lycopene, polyphenols and several other phytochemicals [34]. Moreover, vegetables and fruits were sources of eicosapentaenoic acid (EPA, 20:5 (n–3)) and fish provided docosahexaenoic acid (DHA, 22:6 (n–3)) [33]. The n–6:n–3 ratio of the traditional Cretan diet is estimated to be 2:1 whereas in Europe and the USA the same ratio is 10–20:1 [34]. In other words the traditional Cretan diet was low in saturated fat, high in monounsaturated fat, while providing a variety of bioprotective nutrients. However, is this information appropriately presented by the scientists or communicated to the public via what is called 'Mediterranean diet'?

The Communication of the Traditional Cretan Diet to the Public and Scientific Community

The above-discussed profile of the traditional Cretan diet does not fit completely with the pattern that has been until recently described in scientific journals or communicated to the public. Indeed the traditional Cretan diet, communicated as 'Mediterranean diet', is characterized as a diet with high consumption of olive oil, legumes, fruits, vegetables and cereals (including bread), low consumption of meat and its products and moderate consumption of milk, dairy products and alcohol [18]. The visualization of the above was generated during the International Conference on Diets of the Mediterranean in 1993, and was given a form of a pyramid (fig. 4) [10]. Several years later this graphic was slightly updated including regular physical activity in the recommendations [36].

Table 1. Foods eaten in the traditional Cretan Mediterranean diet [from 32]

	Breakfast	Mid-morning	Lunch	Mid-afternoon	Dinner
Monday	Ksinohontros[1] rusk, orange	Pear	Broad beans, onion, salad (cucumber, tomato, purslane, olives, olive oil), whole-wheat bread, apple, red wine	Walnuts, dry figs	Boiled vegetables Potatoes, olive oil, boiled egg, melon, red wine
Tuesday	Rusk, cheese, apple	Orange	Snails, potatoes and vegetables, salad (tomato, cucumber, onion, olive oil), whole-wheat bread, red wine, longan	Halva[2] (homemade)	Rice with spinach, yogurt, whole wheat bread, longan
Wednesday	Doughnuts (homemade) with honey, apple, herbal tea	Pear	Chickpeas, herring, salad (tomato, cucumber, olives, olive oil), whole-wheat bread, cherries, red wine	Walnuts, dry figs, raki	Stuffed tomatoes, whole wheat bread, salad (tomato, cucumber, onion)
Thursday	Fresh whole milk boiled with ground wheat	Melon	Fish, broad beans (puree), oil, lemon juice, whole-wheat rusk, salad (tomato, cucumber, olives, olive oil, onion), pear, red wine	Halva (homemade)	Lentils, salad (tomato, cucumber, olives, olive oil, onion), apple, red wine, cheese, whole wheat bread

Table 1. (continued)

	Breakfast	Mid-morning	Lunch	Mid-afternoon	Dinner
Friday	Rusk, olives, herbal, tea, apple	Apple	Beans, potatoes, whole-wheat bread, olives, orange	Walnuts, dry figs, raki	Broad beans, artichoke, olive oil, rusk, red wine, melon
Saturday	Milk and whole wheat, melon	Apple	Chicken, okra, potatoes, salad (lettuce, cucumber, olives, olive oil)	Homemade cheese pie, honey, coffee	Boiled vegetables with olive oil, rusk, red wine, melon
Sunday	Homemade cheese pie with honey, melon		Rabbit, pasta, salad (tomato, cucumber, olives, olive oil, onion), rusk, wine, orange	Coffee, halva	Fish, fish soup with vegetables, rusk, red wine apple

[1]Greek foods: Ksinohontros = yogurt, wheat; [2]halva (homemade) = semolina, olive oil, sugar, walnuts.

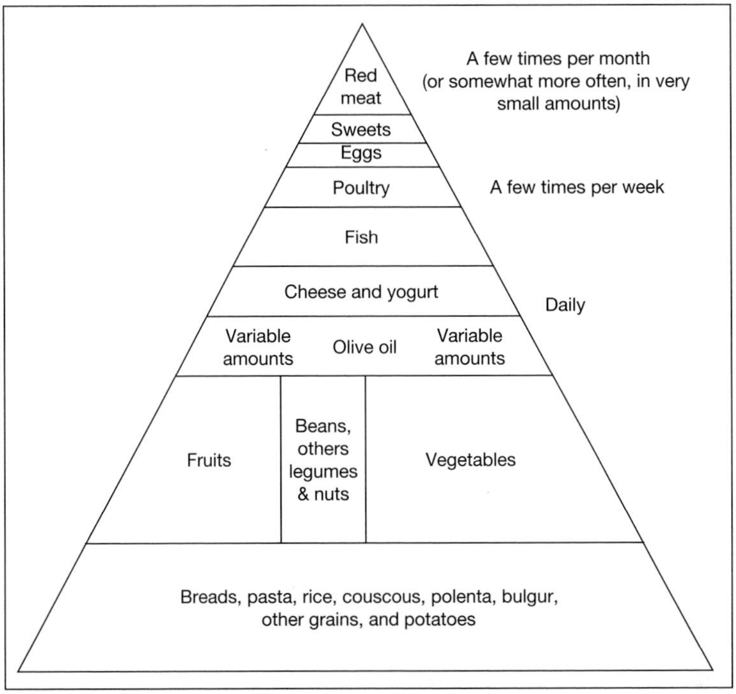

Fig. 4. The Mediterranean diet pyramid of 1993, developed at the International Conference on Diets of the Mediterranean [10].

There is growing evidence that the adoption of a diet conforming to the above norms has beneficial effects on longevity of Mediterranean (e.g. Greeks [37], Spanish [38]) and non-Mediterranean (e.g. Danish [39], Australians [40], Finns [41]) populations, which is accompanied by a reduced mortality rate from coronary heart disease, cardiovascular diseases and some types of cancer [41]. This dietary scheme can also moderate the progression of coronary heart disease as reviewed elsewhere [42], while reducing the risk of peripheral arterial disease in subjects with type 2 diabetes [43]. With any given limitations, the above studies provide some evidence that the communicated Mediterranean diet in the form of the above guidelines, usually assessed with the use of food scores [37, 40, 44], can be transposed to non-Mediterranean cultures in hand with health-boosting outcomes [40, 45].

A question that arises is whether a subject trying to follow these dietary guidelines [10, 36] gets the same nutrients as if he or she followed the traditional Cretan diet. In other words, are these guidelines truly reflecting the nutrient intake of the traditional Cretan diet or do they just mimic a vegetarian diet with moderate meat consumption [46]? Certain evidence has shown that the adoption of such a diet can in many cases [47, 48] but not all [49, 50] decrease LDL or total cholesterol levels and this could possibly explain the lower morbidity and mortality associated with the adopted guidelines. These effects could have been mediated by a decline in the intake of red meat, saturated and trans fatty acids, known to unfavorably influence the lipid profile, a hypothesis which was also proposed by the researchers of the Seven Countries Study [25]. Furthermore, the 25-year follow-up of the Seven Countries Study has indicated that Cretans, although having the same levels of serum cholesterol, enjoy lower rates of morbidity and mortality in comparison with other European countries [51]. In this context it is worth further evaluating the underlying protective components of the traditional Cretan diet, which may not be sufficiently communicated by the Mediterranean pyramid of 1993 and its latest version of 2000 [10, 36].

Indeed, as extensively discussed by Kok and Kromhout [52], the flavonoid and n–3 content of the traditional Cretan diet could be a forgotten missing part of the Cretan longevity puzzle. The analysis of Cretan plants and herbs confirmed their high antioxidant content [31, 53]. It has also been shown that cholesteryl esters of a Cretan population were threefold richer in α-linolenic acid when compared to the Zutphen cohort [54]. It is proposed that the amount of n–3 fatty acids (e.g. α-linolenic acid) was initially underestimated, as a result of inadequate information on wild greens, like purslane, which is the richest source of α-linolenic acid among leafy vegetables that have been until now analyzed [55–57]. With respect to the above biochemical data related to Cretan longevity, it could be claimed that the adherence to the so-called 'Mediterranean guidelines' is not always followed by the traditional Cretan biochemical profile.

The above were further substantiated by the Lyon Diet Heart Study [58]. In this study, 605 patients with myocardial infarction were randomized to the 'Mediterranean diet rich in α-linolenic acid' or a diet similar to the NCEP Step I recommendations [59, 60]. It is remarkable that the nutrient content of the first diet as well as its resulting biochemical effects were comparable to those of the traditional Cretan diet. The 'Mediterranean diet rich in α-linolenic acid' exerted a salutary effect by reducing all-cause mortality by 70% after 5 years of follow-up [58]. Similar effects were observed for cancer mortality (61% reduction at 4 years of follow-up). The traditional risk factors, including total cholesterol, high density lipoprotein cholesterol, low density lipoprotein, triglycerides, glycated hemoglobin and blood pressure were not different between the two groups [58, 61], a finding which is compatible with the 25-year follow-up of the Seven Countries Study [51]. However, changes were detected in the composition of circulating blood lipids and platelet phospholipids, which were favorable for those following the 'Mediterranean diet rich in α-linolenic acid' (lower n–6:n–3 ratio), an effect that has been confirmed by following studies [62]. Moreover, the group conforming to this diet exhibited higher plasma antioxidant vitamins (e.g. α-tocopherol and ascorbic acid) [58]. However, other trials [63] have not demonstrated an association of the current Mediterranean guidelines [10, 36] or simply a diet rich in vegetables and fruits with plasma antioxidant capacity [64] or certain antioxidants [64, 65]. These differences could be attributed to unintentionally neglecting the importance of certain food components of the traditional Cretan diet, such as wild greens and herbs, which are particularly rich in n–3 fatty acids, polyphenols and antioxidants [10, 33].

It is assumed that some communicated approaches of the Mediterranean guidelines [10, 36] may accord only macroscopically with the traditional Cretan profile. From the nutritional point of view, the diet of Cretans may possibly have a different content of various protective substances when compared to a diet, which follows the precepts of the Mediterranean guidelines [10, 36]. Despite the fact that the full bioprotective profile of the traditional Cretan diet has not yet been in all clarified, it is assumed that the consumption of the food groups that 'mimic' the traditional Cretan diet may not provide identical nutrients. It is suggested that certain important foods such as wild green plants, snails, figs, walnuts, etc. have not been appropriately highlighted in the Mediterranean guidelines [10, 23]. Given that attention is directed towards the reinforcement of qualitative aspects of the traditional Cretan diet [11], it is important to take a closer look of the nutrient content of the traditional Cretan food palette.

Fruits, Nuts and Vegetables

Indeed, fruits and vegetables of the Cretan diet are good sources of folate, calcium, minerals, vitamins E and C, glutathione, polyphenols and other antioxidants [33, 66]. Special reference has to be made to some popular fruits, such as figs that Cretans commonly consumed and which are a good source of n–3 fatty acids. A good source of n–3 fatty acids are also walnuts, which were consumed as a snack in the traditional Cretan diet. Another worth mentioning component of the traditional Cretan diet are wild greens, which have a considerable flavonoid and omega–3 content [31, 33, 57]. Furthermore, wild greens consumed in Crete have a higher content of flavonols in comparison with fresh vegetables commonly consumed in Europe, which is important in the context of the association of flavonoids with cardiovascular risk reduction and cancer [67, 68]. It is remarkable that wild greens contain more flavonoids than an equal quantity of red wine or black tea, which for other countries are the main flavonoid sources [26, 69]. The increased light exposure, part of Crete's climate, could also provide an explanatory mechanism [70]. It should also be mentioned that the total phenolic content of chemically analyzed Cretan edible blites was found higher than those of Northern Greece [71]. The polyphenolic and antioxidant content of wild greens is also higher when compared to cultivated vegetables [9, 53]. Moreover, wild edible plants are good sources of α-linolenic acid and contain higher amounts of it than cultivated plants [57]. For example, purslane contains 400 mg of α-linolenic acid in 100 g [72].

Tomatoes are another important component of the traditional Cretan diet and indispensable ingredient of the Greek salad [32]. Intervention studies with different tomato products demonstrate increased protection from oxidative stress, as evaluated by analyzing different markers (e.g. DNA damage and lipid peroxidation) [73, 74]. This protective activity has been often attributed to the tomato antioxidant lycopene [75]. However, tomato products are also sources of other compounds with healthful properties, such as other carotenoids, flavonoids, and vitamin C. From this standpoint, it is noteworthy that co-ingestion of several different foods rich in antioxidants, such as fruits and vegetables, found in abundance in the traditional Cretan diet, can play a significant role in the observed protection from oxidative stress [76].

Meat, Meat Products, Eggs, Fish and Snails

On the one hand, epidemiological studies have shown that red meat consumption is related to a high risk of coronary heart disease and some types of cancer [78, 79]. On the other hand, fish consumption is negatively associated with cardiovascular disease [18]. What is worth mentioning for the traditional Cretan diet is the nutritional content of the animal and animal products consumed. It cannot be ignored that fresh fish, sardines, herrings, seafood as well as game and snails contain EPA and DHA. Moreover, animals of Crete grazed from grass and also ate fruits (like dried figs) and nuts from the ground rather than being fed. In other words, what sheep, chicken and goats consumed was a good source of omega–3 fatty acids (mainly α-linolenic fatty acid), which was also found in their meat and products [34]. That is the reason why meat, cheese, eggs and snails in Crete had a high concentration of n–3 [33]. For instance, the ratio of n–6:n–3 in Cretan eggs has been found to be 1.3, whereas the average USA egg has a ratio of 19.4 [80].

Olive Oil

An inherent characteristic of the traditional Cretan diet in contrast with Northern Europe was the use of olive oil and more particularly extra virgin olive oil [77]. Its nutritional content may provide a possible explanatory mechanism for the longevity of Cretans. In addition to its peculiar fatty acid composition (high proportion of oleic acid), extra virgin olive oil contains a variety of minor components [81]. Indeed, the concentration of the phenolic fraction (up to 800 mg/kg oil) in the oil can predict its stability to oxidation [82]. It is noteworthy that 'extra virgin' olive oil, obtained by squeezing the olive paste and

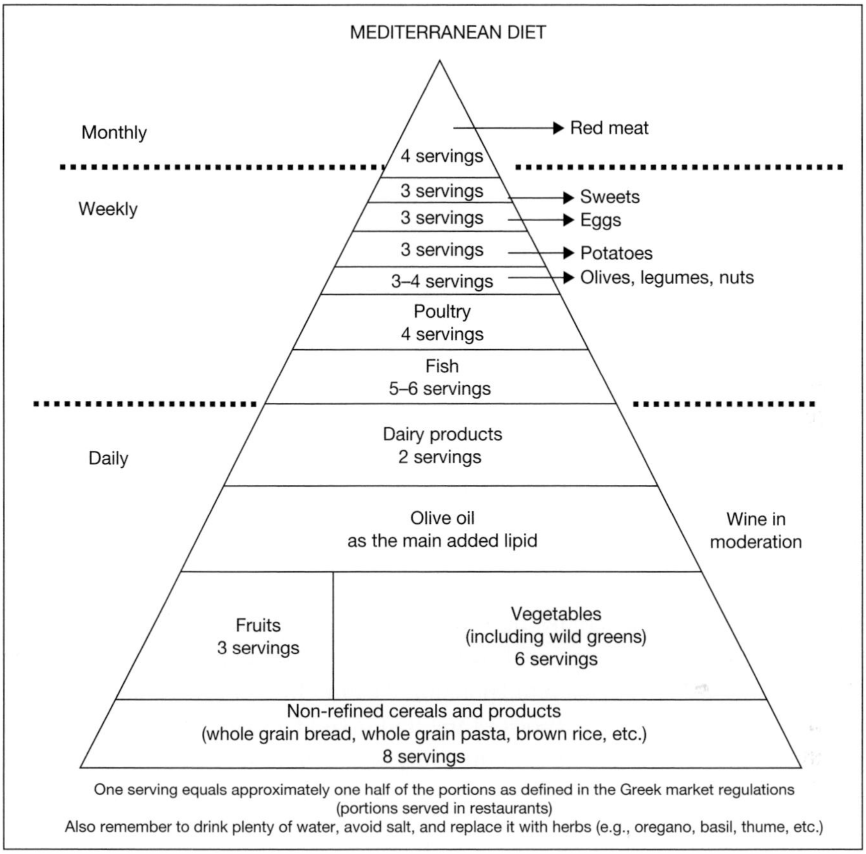

Fig. 5. The Mediterranean pyramid of Greece (Source: Supreme Scientific Health Council, Hellenic Ministry of Health).

with a low concentration of free fatty acids (acidity lower than 1%, as requested by the current regulations), is much richer in phenolic compounds than refined oils, that are obtained by neutralization of acidity from oils that exceeds the given limits and are virtually devoid of phenols. Furthermore, these compounds are responsible for the unique flavor and taste of the extra virgin olive oil. This is related to the fact that olive oil is the only vegetable oil obtained from *whole fruits* rather than from *seeds* and, thus, it retains all the organoleptic properties of olives [82]. In turn, the traditional Cretan diet included the consumption of extra virgin oil which has a unique flavor and taste. In the light of scientific evidence, this dietary scheme rich in extra virgin olive oil can be used as a guiding tool for the formation of nutritional guidelines.

The above particular characteristics of the traditional Cretan are better presented in the proposed Mediterranean pyramid by the Hellenic National Center for Nutrition and Hellenic Ministry of Health [83]. From figure 5 it can be assumed that the Greek Mediterranean diet pyramid is slightly differentiated from the Mediterranean pyramid of 1993 and 2000 [10, 23], but it reflects in a greater extent the features of the Cretan Mediterranean diet. More specifically, this pyramid quantifies the recommendations of different food groups and highlights the consumption of whole grain cereals as well as that of wild greens, while reminding the public to substitute salt for herbs (oregano, basil, thyme, etc.), which have a considerable antioxidant capacity [53]. The fact that a food guide pyramid should promote the consumption of foods rich in antioxidants has been recently proposed in the Fifth Barcelona International Congress on the Mediterranean Diet [84].

Nevertheless, the oversimplified Mediterranean pyramid of 1993 and its latest form of 2000 [10, 23], as a food guidance tool for worldwide consumers, provide some guidance but not as accurate as to reflect the traditional Cretan diet identity and possible health benefits. After the completion of the Lyon Diet Heart Study the research team believed that 'the favourable life expectancy of the Cretans could be largely due to their diet . . . if any Mediterranean diet should be duplicated as a model of healthy diet, it should be the Cretan diet' [85]. On the way to elucidate the potential health effects of micronutrients and non-nutritive substances in foods of the traditional Cretan diet, it is imperative that the efforts of the scientific community with that of media, local communities and agro-food producers are instrumented in order for the message of the protective profile of this form of diet to be appropriately conveyed and reinforced. Despite the increasing knowledge on the potential health benefits of certain components of the traditional Cretan diet, there is still much to be done and until then the only 'effective' message to be conveyed to the public should be based on a diet holistic approach appropriately emphasizing the principles of the traditional Cretan diet.

References

1 Willett WC: Nutritional Epidemiology. New York, Oxford University Press, 1998.

2 Lagiou P, Trichopoulou A, Trichopoulos D: Nutritional epidemiology of cancer: accomplishments and prospects. Proc Nutr Soc 2002;61:217–222.

3 Slattery ML, Boucher KM, Caan BJ, Potter JD, Ma KN: Eating patterns and risk of colon cancer. Am J Epidemiol 1998;148:4–16.

4 Hu FB, Stampfer MJ, Manson JE, Grodstein F, Colditz GA, Speizer FE, Willett WC: Trends in the incidence of coronary heart disease and changes in diet and lifestyle in women. N Engl J Med 2000;343:530–537.

5 Fung TT, Willett WC, Stampfer MJ, Manson JE, Hu FB: Dietary patterns and the risk of coronary heart disease in women. Arch Intern Med 2001;161:1857–1862.

6 Terry P, Hu FB, Hansen H, Wolk A: Prospective study of major dietary patterns and colorectal cancer risk in women. Am J Epidemiol 2001;154:1143–1149.

7 Nielsen N, Hansen J: Breast cancer in Greenland-selected epidemiological, clinical and histological features. J Cancer Res Clin Oncol 1980;98:287–299.

8 Curtis BM, O'Keefe JH Jr: Understanding the Mediterranean diet. Could this be the new 'gold standard' for heart disease prevention? Postgrad Med 2002;112:35–38, 41–45.

9 Trichopoulou A, Lagiou P, Kuper H, Trichopoulos D: Cancer and Mediterranean dietary traditions. Cancer Epidemiol Biomarkers Prev 2000;9:869–873.

10 Willett WC, Sacks F, Trichopoulou A, Drescher G, Ferro-Luzzi A, Helsing E, Trichopoulos D: Mediterranean diet pyramid: a cultural model for healthy eating. Am J Clin Nutr 1995;61: 1402S–1406S.

11 Serra-Majem L, Trichopoulou A, Ngo de la Cruz J, Cervera P, Garcia Alvarez A, La Vecchia C, Lemtouni A, Trichopoulos D: Does the definition of the Mediterranean diet need to be updated? Public Health Nutr 2004;7:927–929.

12 Gjonca A, Bobak M: Albanian paradox, another example of protective effect of Mediterranean lifestyle? Lancet 1997;350:1815–1817.

13 Trichopoulou A, Lagiou P: Healthy traditional Mediterranean diet: an expression of culture, history, and lifestyle. Nutr Rev 1997;55:383–389.

14 Dietary fat consensus statements. Am J Med 2002;113(suppl 9B):5S–8S.

15 Renaud S, Gueguen R: The French paradox and wine drinking. Novartis Found Symp 1998;216: 152–158, 208–222.

16 Sarri KO, Tzanakis NE, Linardakis MK, Mamalakis GD, Kafatos AG: Effects of Greek Orthodox Christian Church fasting on serum lipids and obesity. BMC Public Health 2003;3:16.

17 Sarri KO, Linardakis MK, Bervanaki FN, Tzanakis NE, Kafatos AG: Greek Orthodox fasting rituals: a hidden characteristic of the Mediterranean diet of Crete. Br J Nutr 2004;92:277–284.

18 Trichopoulou A: The Mediterranean diet: definition. Epidemiological aspects and current patterns; in Matalas A, Stavrinos V, Wolinsky I (eds): Mediterranean Diet: Constituents and Health Promotion. Washington, CRC Press, 2001.

19 Hill M, Giacosa A: The Mediterranean diet. Eur J Cancer Prev 1992;1:339–340.

20 Lupo A: Nutrition in general practice in Italy. Am J Clin Nutr 1997;65:1963S–1966S.

21 Keys A, Aravanis C, Sdrin H: The diets of middle-aged men in two rural areas of Greece; in Hartog C, Buzina K, Fidanza K, Keys A, Roine P (eds): Dietary Studies and Epidemiology of Heart Diseases. The Hague, The Stichting, 1968, pp 57–68.

22 Ferro-Luzzi A, Sette S: The Mediterranean diet: an attempt to define its present and past composition. Eur J Clin Nutr 1989;43(suppl 2):13–29.

23 Simopoulos AP, Visioli F (eds): Mediterranean Diets. World Rev Nutr Diet. Basel, Karger, 2000, vol 87, pp 1–184.

24 Albaugh L: Crete: A Case Study of an Undeveloped Area. Princeton, Princeton University Press, 1953.

25 Keys A: Coronary heart disease in seven countries. Circulation 1970;41:1–211.

26 Hertog MG, Kromhout D, Aravanis C, Blackburn H, Buzina R, Fidanza F, Giampaoli S, Jansen A, Menotti A, Nedeljkovic S, et al: Flavonoid intake and long-term risk of coronary heart disease and cancer in the seven countries study. Arch Intern Med 1995;155:381–386.

27 Kromhout D: Serum cholesterol in cross-cultural perspective. The Seven Countries Study. Acta Cardiol 1999;54:155–158.

28 Dontas AS, Menotti A, Aravanis C, Ioannidis P, Seccareccia F: Comparative total mortality in 25 years in Italian and Greek middle-aged rural men. J Epidemiol Community Health 1998;52: 638–644.

29 Simini B: Serge Renaud: from French paradox to Cretan miracle. Lancet 2000;355:48.

30 Nestle M: Mediterranean diets: historical and research overview. Am J Clin Nutr 1995;61: 1313S–1320S.

31 Trichopoulou A, Hollman P, Chanialides C, Foufa B, Kaloudis, et al: Nutritional composition and flavonoid content of edible wild greens and green pies: a potential rich source of antioxidant nutrients in the Mediterranean diet. Food Chem 2000;70:319–323.

32 Kafatos A, Verhagen H, Moschandreas J, Apostolaki I, Van Westerop JJ: Mediterranean diet of Crete: foods and nutrient content. J Am Diet Assoc 2000;100:1487–1493.

33 Simopoulos AP: The Mediterranean diets: what is so special about the diet of Greece? The scientific evidence. J Nutr 2001;131:3065S–3073S.
34 Simopoulos AP, Sidossis L: What is so special about the traditional diet of Greece. The scientific evidence; in Simopoulos AP (ed): Mediterranean Diets. World Rev Nutr Diet. Basel, Karger, 2000, pp 24–42.
35 Christakis G, Severinghaus EL, Maldonado Z, Kafatos FC, Hashim SA: Crete: a study in the metabolic epidemiology of Coronary Heart Disease. Am J Cardiol 1965;15:320–32.
36 The Traditional Healthy Mediterranean Diet Pyramid. Oldways Preservation and Exchange Trust, 2000 www.oldwayspt.org
37 Trichopoulou A, Costacou T, Bamia C, Trichopoulos D: Adherence to a Mediterranean diet and survival in a Greek population. N Engl J Med 2003;348:2599–2608.
38 Lasheras C, Fernandez S, Patterson AM: Mediterranean diet and age with respect to overall survival in institutionalized, nonsmoking elderly people. Am J Clin Nutr 2000;71:987–992.
39 Osler M, Schroll M: Diet and mortality in a cohort of elderly people in a north European community. Int J Epidemiol 1997;26:155–159.
40 Kouris-Blazos A, Gnardellis C, Wahlqvist ML, Trichopoulos D, Lukito W, Trichopoulou A: Are the advantages of the Mediterranean diet transferable to other populations? A cohort study in Melbourne, Australia. Br J Nutr 1999;82:57–61.
41 Knoops KT, de Groot LC, Kromhout D, Perrin AE, Moreiras-Varela O, Menotti A, van Staveren WA: Mediterranean diet, lifestyle factors, and 10-year mortality in elderly European men and women: the HALE project. JAMA 2004;292:1433–1439.
42 Panagiotakos DB, Pitsavos C, Polychronopoulos E, Chrysohoou C, Zampelas A, Trichopoulou A: Can a Mediterranean diet moderate the development and clinical progression of coronary heart disease? A systematic review. Med Sci Monit 2004;10:RA193–198.
43 Ciccarone E, Di Castelnuovo A, Salcuni M, Siani A, Giacco A, Donati MB, De Gaetano G, Capani F, Iacoviello L: A high-score Mediterranean dietary pattern is associated with a reduced risk of peripheral arterial disease in Italian patients with Type 2 diabetes. J Thromb Haemost 2003;1: 1744–1752.
44 Osler M, Heitmann BL, Gerdes LU, Jorgensen LM, Schroll M: Dietary patterns and mortality in Danish men and women: a prospective observational study. Br J Nutr 2001;85:219–225.
45 Biesalski HK: Editorial. Eur J Nutr 2004;43(suppl 1):I/1.
46 Coulston AM: The role of dietary fats in plant-based diets. Am J Clin Nutr 1999;70:512S–515S.
47 Ehnholm C, Huttunen JK, Pietinen P, Leino U, Mutanen M, Kostiainen E, Pikkarainen J, Dougherty R, Iacono J, Puska P: Effect of diet on serum lipoproteins in a population with a high risk of coronary heart disease. N Engl J Med 1982;307:850–855.
48 Goulet J, Lamarche B, Nadeau G, Lemieux S: Effect of a nutritional intervention promoting the Mediterranean food pattern on plasma lipids, lipoproteins and body weight in healthy French-Canadian women. Atherosclerosis 2003;170:115–124.
49 Toobert DJ, Glasgow RE, Strycker LA, Barrera M Jr, Radcliffe JL, Wander RC, Bagdade JD: Biologic and quality-of-life outcomes from the Mediterranean Lifestyle Program: a randomized clinical trial. Diabetes Care 2003;26:2288–2293.
50 Mezzano D, Leighton F, Strobel P, Martinez C, Marshall G, Cuevas A, Castillo O, Panes O, Munoz B, Rozowski J, Pereira J: Mediterranean diet, but not red wine, is associated with beneficial changes in primary haemostasis. Eur J Clin Nutr 2003;57:439–446.
51 Verschuren WM, Jacobs DR, Bloemberg BP, Kromhout D, Menotti A, Aravanis C, Blackburn H, Buzina R, Dontas AS, Fidanza F, et al: Serum total cholesterol and long-term coronary heart disease mortality in different cultures. Twenty-five-year follow-up of the Seven Countries Study. JAMA 1995;274:131–136.
52 Kok FJ, Kromhout D: Atherosclerosis-epidemiological studies on the health effects of a Mediterranean diet. Eur J Nutr 2004;43(suppl 1):I/2–5.
53 Lionis C, Faresjo A, Skoula M, Kapsokefalou M, Faresjo T: Antioxidant effects of herbs in Crete. Lancet 1998;352:1987–1988.
54 Sandker GW, Kromhout D, Aravanis C, Bloemberg BP, Mensink RP, Karalias N, Katan MB: Serum cholesteryl ester fatty acids and their relation with serum lipids in elderly men in Crete and The Netherlands. Eur J Clin Nutr 1993;47:201–208.

55 Stampfer MJ, Malinow MR, Willett WC, Newcomer LM, Upson B, Ullmann D, Tishler PV, Hennekens CH: A prospective study of plasma homocyst(e)ine and risk of myocardial infarction in US physicians. JAMA 1992;268:877–881.
56 Simopoulos AP, Norman HA, Gillaspy JE: Purslane in human nutrition and its potential for world agriculture. World Rev Nutr Diet 1995;77:47–74.
57 Simopoulos AP: Omega–3 fatty acids and antioxidants in edible wild plants. Biol Res 2004;37: 263–277.
58 De Lorgeril M, Renaud S, Mamelle N, Salen P, Martin JL, Monjaud I, Guidollet J, Touboul P, Delaye J: Mediterranean α-linolenic acid-rich diet in secondary prevention of coronary heart disease. Lancet 1994;343:1454–1459.
59 Rivellese AA: Monounsaturated and marine omega–3 fatty acids in NIDDM patients. Ann NY Acad Sci 1997;827:302–309.
60 De Lorgeril M, Salen P, Caillat-Vallet E, Hanauer MT, Barthelemy JC, Mamelle N: Control of bias in dietary trial to prevent coronary recurrences: The Lyon Diet Heart Study. Eur J Clin Nutr 1997;51:116–122.
61 De Lorgeril M, Salen P, Martin JL, Monjaud I, Delaye J, Mamelle N: Mediterranean diet, traditional risk factors, and the rate of cardiovascular complications after myocardial infarction: final report of the Lyon Diet Heart Study. Circulation 1999;99:779–785.
62 Urquiaga I, Guasch V, Marshall G, San Martin A, Castillo O, Rozowski J, Leighton F: Effect of Mediterranean and Occidental diets, and red wine, on plasma fatty acids in humans. An intervention study. Biol Res 2004;37:253–261.
63 Hagfors L, Leanderson P, Skoldstam L, Andersson J, Johansson G: Antioxidant intake, plasma antioxidants and oxidative stress in a randomized, controlled, parallel, Mediterranean dietary intervention study on patients with rheumatoid arthritis. Nutr J 2003;2:5.
64 Record IR, Dreosti IE, McInerney JK: Changes in plasma antioxidant status following consumption of diets high or low in fruit and vegetables or following dietary supplementation with an antioxidant mixture. Br J Nutr 2001;85:459–464.
65 John JH, Ziebland S, Yudkin P, Roe LS, Neil HA: Effects of fruit and vegetable consumption on plasma antioxidant concentrations and blood pressure: a randomised controlled trial. Lancet 2002;359:1969–1974.
66 Nanos G, Grasopoulos D: Fruits, vegetables, legumes and grains; in Matalas A, Zampelas A, Stavrinos V, Wolinsky I (eds): Mediterranean Diet: Constituents and Health Promotion. Washington, CRC Press, 2001.
67 Knekt P, Jarvinen R, Reunanen A, Maatela J: Flavonoid intake and coronary mortality in Finland: a cohort study. BMJ 1996;312:478–481.
68 Geleijnse JM, Launer LJ, Van der Kuip DA, Hofman A, Witteman JC: Inverse association of tea and flavonoid intakes with incident myocardial infarction: the Rotterdam Study. Am J Clin Nutr 2002;75:880–886.
69 Manach C, Scalbert A, Morand C, Remesy C, Jimenez L: Polyphenols: food sources and bioavailability. Am J Clin Nutr 2004;79:727–747.
70 Stewart AJ, Bozonnet S, Mullen W, Jenkins GI, Lean ME, Crozier A: Occurrence of flavonols in tomatoes and tomato-based products. J Agric Food Chem 2000;48:2663–2669.
71 Greek Food Composition Tables. University of Crete and Technological Educational Institution of Thessaloniki, June 2001.
72 Simopoulos AP, Norman HA, Gillaspy JE, Duke JA: Common purslane: a source of omega–3 fatty acids and antioxidants. J Am Coll Nutr 1992;11:374–382.
73 Porrini M, Riso P, Brusamolino A, Berti C, Guarnieri S, Visioli F: Daily intake of a formulated tomato drink affects carotenoid plasma and lymphocyte concentrations and improves cellular antioxidant protection. Br J Nutr 2005;93:93–99.
74 Visioli F, Riso P, Grande S, Galli C, Porrini M: Protective activity of tomato products on in vivo markers of lipid oxidation. Eur J Nutr 2003;42:201–206.
75 Riso P, Visioli F, Erba D, Testolin G, Porrini M: Lycopene and vitamin C concentrations increase in plasma and lymphocytes after tomato intake. Effects on cellular antioxidant protection. Eur J Clin Nutr 2004;58:1350–1358.

76 Visioli F, Grande S, Bogani P, Galli C: The role of antioxidants in the Mediterranean diets: focus on cancer. Eur J Cancer Prev 2004;13:337–343.
77 Kromhout D, Keys A, Aravanis C, Buzina R, Fidanza F, Giampaoli S, Jansen A, Menotti A, Nedeljkovic S, Pekkarinen M, et al: Food consumption patterns in the 1960s in seven countries. Am J Clin Nutr 1989;49:889–894.
78 Kushi LH, Lenart EB, Willett WC: Health implications of Mediterranean diets in light of contemporary knowledge. 2. Meat, wine, fats, and oils. Am J Clin Nutr 1995;61:1416S–1427S.
79 Food, Nutrition and the Prevention of Cancer: A Global Perspective. Washington, World Cancer Research Fund and American Institute for Cancer Prevention, 1997.
80 Simopoulos AP, Salem N Jr: Egg yolk as a source of long-chain polyunsaturated fatty acids in infant feeding. Am J Clin Nutr 1992;55:411–414.
81 Visioli F, Galli C: The effect of minor constituents of olive oil on cardiovascular disease: new findings. Nutr Rev 1998;56:142–147.
82 Visioli F, Poli A, Gall C: Antioxidant and other biological activities of phenols from olives and olive oil. Med Res Rev 2002;22:65–75.
83 Supreme Scientific Health Council MoHaWoG: Dietary guidelines for adults in Greece. Arch Hellenic Med 1999;16:516–524.
84 Basulto J, Rigolfas R, Lizarraga A: Piramide de la alimentacion rica en antioxidantes. Fifth Barcelona International Congress on the Mediterranean Diet, Barcelona 2004.
85 Renaud S, de Lorgeril M, Delaye J, Guidollet J, Jacquard F, Mamelle N, Martin JL, Monjaud I, Salen P, Toubol P: Cretan Mediterranean diet for prevention of coronary heart disease. Am J Clin Nutr 1995;61:1360S–1367S.

Yannis Manios
Department of Nutrition and Dietetics, Harokopio University
GR–17671 Athens (Greece)
Tel. +30 210 9549156, Fax +30 210 9549141, E-Mail manios@hua.gr

Heinrich M, Müller WE, Galli C (eds): Local Mediterranean Food Plants and Nutraceuticals. Forum Nutr. Basel, Karger, 2006, vol 59, pp 171–181

Development of Bioactive Substances for Functional Foods – Scientific and Other Aspects

Peter Weber, Beat Flühmann, Manfred Eggersdorfer

DSM Nutritional Products Ltd, Human Nutrition & Health, Basel, Switzerland

Abstract

The discovery and development of bioactive substances and their use in the manufacture of food products has the potential to contribute to the optimal health of populations and in reducing the risk of chronic disease. In assessing the efficacy of these substances in man, the concept of biomarkers could play a key role. However, it is the validation of such biomarkers which is crucial to the eventual availability of foods containing bioactive substances. Determining the safety and efficacy of these substances, and in particular in establishing claims, requires close cooperation between industry, regulatory authorities and the scientific community.

Changes in Nutrition – A Historical Review

The fundamental role of nutrition is to ensure an adequate intake of macro- and micronutrients, enabling the efficient operation of metabolic processes. This postulate has led to the development of recommendations on nutrient intake (table 1). As long ago as the 1940s, the first recommendations for selected nutrients were published in the USA [1] with the aim of avoiding nutrient deficiency, classically manifested for example as scurvy in the case of vitamin C deficiency.

It was subsequently recognized that, in addition to the problem of inadequate intake, certain nutrients can also be consumed in excess. For example, in the 1960s there was an accumulation of scientific evidence that excessive fat intake could cause a rise in blood lipids. The presence of elevated blood lipids, or hyperlipidemia, in turn represents an important risk factor in the development

Table 1. Changes in nutrition – historical review

1940s	Recommendations for selected nutrients aimed at avoiding disease due to inadequate intake (e.g. proteins, minerals, vitamins)
1960s	Excessive intake of certain nutrients can also lead to disease (e.g. fat/cholesterol & atherosclerosis)
1980s	Essential nutrients have a health effect in addition to their essentiality (e.g. vitamin E & cardiovascular disease)
2000	Non-essential nutrients have a health effect (e.g. soya & cardiovascular disease)

of atherosclerosis, as has been subsequently proven [2]. An excess of certain nutrients may thus also have a lasting and significant health impact, including the onset of clinical disease.

This relatively simple concept of inadequacy versus excess was later refined. In the 1980s, there was an increase in the number of scientific reports suggesting that essential nutrients play a part in reducing risk factors for (chronic) disease. One example of this is the role of antioxidant vitamins in the prevention of cardiovascular disease [3].

Finally, at the end of the 1990s, research in nutritional science turned increasingly to the role of non-essential nutrients in human health. Evidence supporting the potential role of non-essential nutrients in reducing the risk of chronic diseases continues to accumulate. The US Food and Drug Administration (FDA) recently even authorized a further health claim to the effect that soy protein may reduce the risk of cardiovascular disease [4].

The realization that nutrition not only serves energy and nutrient requirements, but can also have a significant additional important impact on health, has led to the development of foods with a health-promoting effect.

Foods with a Health-Promoting Effect

What do we mean by foods with a health-promoting effect? Neither the scientific literature nor legislation provides a clear definition. Terms used as synonyms include 'functional foods', 'designer foods', 'pharma foods' and 'agromedical foods', to name but a few. The term 'nutraceuticals' refers to the bioactive substances contained in such foods, and reminds us that these are dietary ingredients, thus distinguishing them from pharmaceuticals [5]. The complexity of this new field of nutrition is underlined by the fact that we not only lack a generally accepted term but have so far failed to agree on a common definition.

Table 2. Some examples of bioactive nutrients that may contribute to improved health

Substance	Occurrence	Health effect
Folic acid	→ vegetables	→ spina bifida
PUFA	→ fish oil	→ cardiovascular
Lutein	→ spinach	→ visual impairment
Lycopene	→ tomatoes	→ prostate
Zeaxanthin	→ maize	→ visual impairment

In the German-speaking world, the Food and Consumer Safety Division of the Swiss Federal Office of Public Health has proposed the following definition of functional foods, which broadly matches that of other groups and authors [6]: '. . . foods with a specific additional benefit that goes beyond the nutritional benefit of the nutrients they contain'. In the USA, the Food and Nutrition Board (FNB) of the National Academy of Sciences has defined functional food as 'a food that encompasses potentially healthful products, including any modified food or food ingredient that may provide a health benefit beyond that of the nutritional nutrients it contains' [7]. This definition certainly describes very well the role of bioactive ingredients in foods with a health-promoting effect. Table 2 gives some examples of nutrients with demonstrated health benefits that go beyond their physiological effects.

In simplified terms the development of bioactive ingredients for foods with health-promoting effects can be divided into the following steps, which are discussed in detail below: identification; safety testing, and efficacy testing.

Identification of Bioactive Ingredients for Functional Foods

The most common procedure for identifying bioactive substances is selection based on data from the available literature. If, however, the objective is the discovery of large numbers of new bioactive substances, then this approach has considerable limitations. In this respect, the technology of high throughput screening (HTS), which has been developed in recent years in pharmacological research [8], offers new opportunities for nutritional research. A modified HTS concept for identifying new bioactive substances suitable for developing functional foods is presented below.

The first – and probably most demanding – step is to identify a molecular mechanism of action by which the sought bioactive substance exerts its effect. Ideally the mechanism of action should be validated in man: that is, a substance

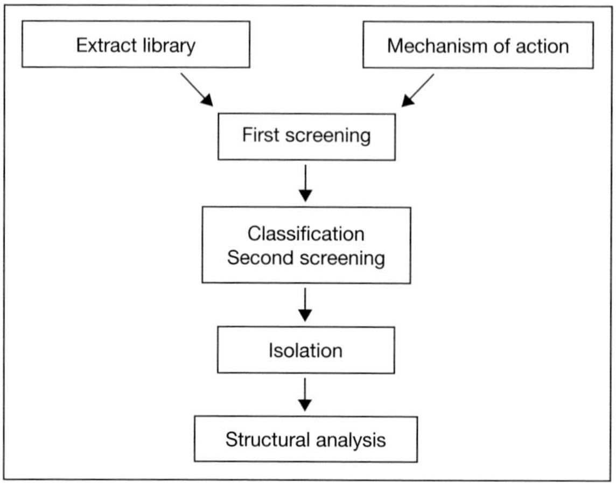

Fig. 1. Flowchart of the screening programme for naturally occurring bioactive substances.

must be identified that leads to the desired health-promoting effect(s) by interacting with the target molecule in the human body. This effect must be measurable by suitable biomarkers and/or endpoints in clinical studies. The HTS process can be broken down into three basic steps, described in more detail below: (i) the establishment of an extract library; (ii) in vitro testing of these extracts to identify 'hot fractions', and (iii) classification of the bioactive substances in these 'hot fractions', including detailed structural analysis (fig. 1).

An extract library with a large chemical diversity of test substances is a precondition for successful HTS. Substances that are part of the food chain – such as extracts of foods from plants, microorganisms, fungi, animals and their combinations – can be used to establish a library of this type. The complex mixture of substances in the extracts may lead to overlapping biological and physicochemical effects, such that active substances remain undetected or that combinations of components produce false-positive results. To minimize these effects, the individual extracts can be separated into further fractions with the aid of high-performance liquid chromatography.

Using HTS, 100,000 or more samples a day can be tested for activity in an in vitro test. In other words, HTS is not the bottleneck in naturally-derived compound screening. The real challenge is the unknown composition and quantity of the complex mixtures in the extract fractions under test. To avoid the time-consuming isolation of false-positive components, great attention must be paid to the quality, and especially the dynamics over a wide concentration range, of the test systems used in HTS.

The next step in naturally-derived compound screening is the rapid and efficient classification of biological activities and the chemical structures on which they are based. Since many substances are present in a variety of extracts, it is vital to classify them as early in the process as possible. The identification of known, already isolated, substances in order to exclude them allows attention to be focused on promising new candidates, and is therefore essential. Thanks to the combination of liquid chromatography and accurate determination of molecular mass by mass spectrometry [9], it is possible to predict the sum formula of a large number of components of an extract and to compare them with substances stored in chemical databases [10]. Many components can be identified by combining the empirical formula with the UV spectrum and taxonomy of the plant extract in question. The remaining unknown components are classified on the basis of the exact mass number, retention time and UV spectrum. By integrating these physical properties with the biological activities identified concurrently by HTS, it is possible to focus on the isolation and detailed structural analysis of promising new substances using nuclear magnetic resonance spectroscopy.

The bioactive substances identified by HTS are then subjected to a series of further in vitro tests, such as bioavailability, stability and solubility, before proceeding to in vivo testing in animals and eventually in humans. Where molecular biological, physiological and clinical knowledge is closely integrated with the technological capabilities of HTS, computing and natural-product chemistry, new bioactive substances can be efficiently identified to develop foods with a health-promoting effect. However, as we shall see later, interdisciplinary collaboration is essential to the success of this approach.

Safety Testing of Bioactive Substances

The essential prerequisite for any bioactive substance intended for use in food is that its safety be established by generally accepted criteria. The number and scope of tests will naturally vary with different classes of substance, but in simplified terms safety testing consists of in vitro tests, experiments on animals and the performance of human studies. These studies are performed in accordance with recommendations and guidelines established both by national legislation and international bodies such as the WHO [11]. A further important element of the safety program is the obtention of pharmacological and pharmacodynamic data which characterizes the activity profiles of these bioactive substances in man. This information forms the basis for planning and performing human efficacy studies. It is worth emphasizing that all human studies – whether in the context of safety testing or the efficacy studies described below – must comply with the ICH Guideline for Good Clinical Practice [12].

Efficacy Testing of Bioactive Substances

Most naturally occurring bioactive substances are not essential to the human body. This means that no known typical signs or symptoms appear if these substances are not regularly consumed. For example, while inadequate vitamin C intake leads after a time to the classic deficiency syndrome of scurvy, such syndromes do not exist for most naturally occurring substances with a postulated health-promoting potential, such as flavonoids, polyphenols and phytoestrogens. This means that a precondition for adding these non-essential bioactive substances to foods is proof of their efficacy in terms of a defined health marker, such as an established biomarker.

Proving that a substance that occurs naturally in the food chain has a defined health-promoting effect confronts scientists with a considerable challenge. The effects of these naturally occurring bioactive substances can only be preventive, or more accurately they tend to reduce established risk factors for chronic disease. Thus, the effects of these naturally occurring bioactive compounds are fundamentally distinct from curative drugs. This means that the effect of these substances on the human body may be minor over relatively short periods – months or a few years – but can contribute significantly to health when they are consumed throughout life as part of the daily diet.

It is also known that many chronic human illnesses begin to develop early in life, but are only manifested – that is, impair individual health – very much later. This time lag between the start of an illness and individually observable symptoms applies to cardiovascular disease, osteoporosis and many forms of cancer, and may last as long as several decades. Atherosclerotic changes can be found in situ in the blood vessels of adolescents and young adults, or even earlier [13]. However, clinical manifestations such as myocardial infarction, stroke or intermittent claudication will generally appear only two or three decades later. The same is true of osteoporosis, which results in bone weakness with an increased risk of fractures. Bone is strongest between the ages of 20 and 30. Afterwards – in women earlier than in men – there is a continuous decline in bone strength, many people suffering fractures between the ages of 50 and 80. Similarly, in malignant diseases of the stomach and prostate, up to 20 years may elapse between the initiation – that is, the appearance of the first malignant cell – and clinical manifestation of the cancer. The list of examples is by no means exhaustive.

Investigating preventive properties in these naturally occurring bioactive substances, when the effect may only be moderate, thus poses a serious dilemma in nutritional research. To facilitate discussion of this problem, we should first briefly examine what scientific possibilities exist for demonstrating the efficacy of these substances.

Table 3. Definitions of biomarker, clinical endpoint and surrogate endpoint

Biomarker A characteristic that is objectively measured and evaluated as an indicator of normal biological processes, pathogenic processes, or pharmacologic responses to a therapeutic intervention
Clinical endpoint A characteristic or variable that reflects how a patient feels, functions, or survives
Surrogate endpoint A biomarker that is intended to substitute for a clinical endpoint

As briefly mentioned earlier, the various categories of scientific efficacy tests comprise in vitro tests (molecular and cellular assays), animal experiments and human studies, the latter being subdivided into epidemiological and intervention studies. Often epidemiological studies – for example, to investigate the relationship between diet and the occurrence of a disease in a relatively large population – lead to a hypothesis, which is investigated further in vitro and in animal experiments. Intervention studies may then be performed to demonstrate efficacy in humans taking a specific amount of a bioactive substance each day under precisely defined and controlled conditions.

The Concept of Biomarkers

The measurements used may be either intermediate endpoints or clinical endpoints. Clinical endpoints, such as myocardial infarction, fractures, or the occurrence or course of cancers, are typically used to investigate the efficacy of drugs, that is, for therapeutic purposes. However, for the reasons discussed above, clinical endpoints appear less suitable for studying the efficacy of bioactive substances; that is, for prevention. We therefore need to take a closer look at the concept of biomarkers and their role in demonstrating the efficacy of bioactive substances.

Table 3 provides the definitions of 'biomarker', 'surrogate marker' and 'clinical endpoint' proposed by Zeger [14] at an NIH workshop on biomarkers and surrogate endpoints. A surrogate endpoint – that is, a biomarker capable of replacing a clinical endpoint – thus appears to be the appropriate parameter for demonstrating the efficacy of naturally occurring bioactive substances in human intervention studies, provided it meets certain criteria [15]: (i) it must display good specificity and sensitivity; (ii) it must be standardized and validated; (iii) it must be non-invasive or, at most, minimally invasive, and (iv) it must be inexpensive if it is to be widely used.

Table 4. Examples of biomarkers and clinical endpoints

Cardiovascular disease
Biomarker: serum cholesterol, blood pressure, etc.
Clinical endpoint: myocardial infarction, stroke
Osteoporosis
Biomarker: markers of bone formation, bone resorption and bone density
Clinical endpoint: bone fractures
Cancer, e.g. prostate
Biomarker: PSA?
Clinical endpoint: prostatic neoplasm or metastases

With these provisos, the biomarker concept could be an essential element, not only in the efficacy testing of bioactive substances but in nutritional science in general. Using biomarkers, the moderate effects of natural bioactive ingredients on the human body could be studied without excessive strain on resources. Most biomarkers are factors that signal the risk of developing a chronic disease at an early stage. They therefore appear to be suitable parameters for investigating the preventive effect of bioactive substances, given that the goal of improved nutrition is to reduce risk factors and hence the probability of developing chronic disease. Some examples of biomarkers are shown in table 4.

Of course, the biomarker concept also has its limitations. A single biomarker will almost certainly never be ideal. Rather, a series of biomarkers will probably be required to examine the effect of a bioactive substance on human health. With this biomarker profile, it will be possible to categorize the effect of a bioactive substance on risk factors for chronic diseases or other functions of the human body.

Lastly, it is important to note that in assessing the efficacy of a bioactive substance, the results of in vitro tests, animal experiments and epidemiological and intervention studies in man must of course be considered as a whole.

Bioactive Substances – Market Opportunities and Challenges

The aim of identifying bioactive substances for incorporation in foods with a health-promoting effect is ultimately to market these substances in foods. Certain safety and efficacy data are required for this purpose, as discussed above. However, the amount of data required is defined by a very dynamic environment. The individual players in this environment are university scientists, regulatory authorities, industry and the consumer (associations), all of which

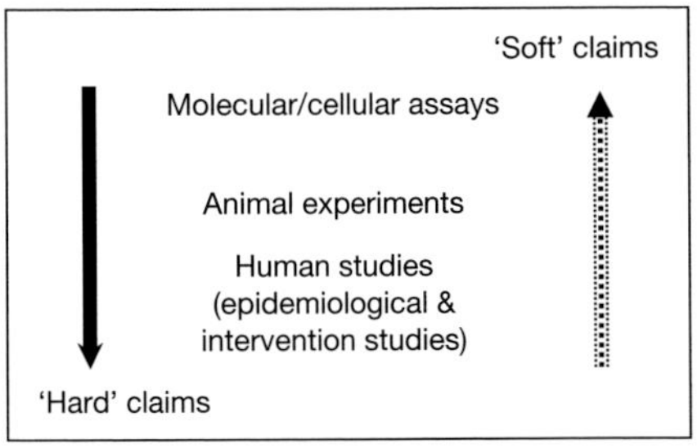

Fig. 2. Scientific data and claims.

have differing expectations and requirements of bioactive substances that may be marketed in foods.

For university scientists the development of bioactive substances as food ingredients offers an exciting research field. The biomarker concept, in particular, represents a golden opportunity to rapidly advance knowledge in nutritional science. An important precondition for this is validation of the biomarkers. This requires close collaboration between scientists in universities, regulatory authorities and industry. Apart from safety, a very important factor in marketing these bioactive substances or foods that contain them is the question of what these substances 'do' and what can be said about them; that is, what claims can be made. Defining this will require dynamic interaction between industry, the regulatory authorities and scientific community.

It would be helpful if these parties could reach a consensus on what efficacy data are needed to make a claim. It should also be possible for the wealth of data accumulated in developing these bioactive substances to be converted into claims in a 'graduated manner'. Differentiation between 'soft' and 'hard' claims could be useful in this respect. Epidemiological data in conjunction with in vitro and animal data could be considered 'soft claims', while results from human intervention studies would be classed as 'hard claims' (fig. 2). Differentiation of the scientific evidence in this way could contribute to more transparency for consumers. The development of bioactive substances that lead to healthier foods should also not be overburdened by demanding a proof of efficacy equal to that required for pharmaceuticals. In this area regulatory uncertainty undoubtedly exists, and should be addressed as part of the dialogue among scientists in universities, regulatory authorities and industry. The aim should be to ensure that knowledge in the nutritional field can be exploited in food production as quickly as possible to the benefit of populations. On the other hand, there should be enough flexibility to allow for changes as soon as better data are available. The sensible use of adequate and validated biomarkers, for example, could help to ensure that discoveries in the nutritional field can be implemented with greater speed, and so contribute to optimized nutrition.

Apart from the tests discussed above in *Safety Testing of Bioactive Substances*, deliberations on the safety of bioactive substances or foods that contain them should take account of a further aspect, namely whether and to what extent these bioactive substances have already been present in the diet (history of safe use and exposure rate). It is conceivable that different levels of safety testing could be necessary for bioactive substances that exert health-promoting effects at doses only slightly above previous exposure ('physiological effect') and those that achieve effects relevant to health only with an intake many times the normal exposure ('pharmacological effect').

Concluding Remarks

The biomarker concept could represent a key element of nutritional science in the development of naturally occurring bioactive substances. The discovery and availability of such substances could contribute greatly to preserving health, especially in populations whose average age is predicted to rise dramatically because of its demographic structure. This fact should not be underestimated, given that healthcare costs are exploding, and any opportunity to reduce the pressure on public health resources should be seized. Finally, to facilitate the marketing of such substances within an acceptable time frame, it is essential that scientists, regulatory authorities and food industry representatives work closely together.

References

1 Food and Nutrition Board: Recommended Dietary Allowances. Washington, National Research Council, January 1943.

2 Assmann G, Schulte H: The Prospective Cardiovascular Münster (PROCAM) Study: prevalence of hyperlipidemia in persons with hypertension and/or diabetes mellitus and the relationship to coronary heart disease. Am Heart J 1988;116:1713–1724.

3 Weber P, Bendich A, Machlin LJ: Vitamin E and human health – rationale for determining recommended intake levels. Nutrition 1997;13:450–460.

4 Food Labeling: Health Claims; Soy Protein and Coronary Heart Disease. Federal Register: Nov 10, 1998;63:#217 (Docket No 98P-0683).

5 Groenveld M: Functional foods: definition and status in food law (in German). Ernähr Umsch 1998;48:156–161.

6 Schweizer BGA, Bulletin 26 of 26th June 2000:497.

7 Milner JA: Functional foods: the US perspective. Am J Clin Nutr 2000;71(suppl):1654S–1659S.

8 Watt A, Morrison D: Strategic and technical challenges for drug discovery. Drug Discov Today 2001;6:290–292.

9 Recently achieved in the form of High-Performance Liquid Chromatography ElectroSpray Ionisation Time of Flight Mass Spectrometer (HPLC-ESI-TOF-MS).

10 Strege MA: High-performance liquid chromatographic-electrospray ionization mass spectrometric analyses for the integration of natural products with modern high-throughput screening. J Chromatogr B Biomed Sci Appl 1999;725:67–78.
11 World Health Organization. Environmental Health Criteria: Principles and Methods for Evaluating the Toxicity of Chemicals. Geneva, WHO, 1986.
12 European Agency for the Evaluation of Medicinal Products, Human Medicines Evaluation Unit. ICH Topic E6. London, Guidelines for Good Clinical Practice, 1997.
13 Stary HC: Macrophages, macrophage foam cells, and eccentric intimal thickening in the coronary arteries of young children. Atherosclerosis 1987;64:91–108.
14 Zeger SL: Biomarkers and Surrogate Endpoints. Advancing Clinical Research and Applications. NIH & FDA Symposium, April 15–16, 1999.
15 Weber P: Role of biomarkers in nutritional science and industry – a comment. Br J Nutr 2001;86(suppl 1):S93–S95.

Dr. Beat Flühmann
DSM Nutritional Products Ltd, Human Nutrition & Health
PO Box 3255, CH–4002 Basel (Switzerland)
Tel. +41 61 6882371, Fax +41 61 6889684, E-Mail beat.fluehmann@dsm.com

Author Index

Subject Index